CONCRETE ABSTRACT ALGEBRA
From Numbers to Gröbner Bases

Concrete Abstract Algebra develops the theory of abstract algebra from numbers to Gröbner bases, whilst taking in all the usual material of a traditional introductory course. In addition there is a rich supply of topics such as cryptography, factoring algorithms for integers, quadratic residues, finite fields, factoring algorithms for polynomials and systems of non-linear equations. A special feature is that Gröbner bases do not appear as an isolated example. They are fully integrated as a subject that can be taught successfully in an undergraduate context.

Lauritzen's approach to teaching abstract algebra is based on an extensive use of examples, applications and exercises. The basic philosophy is that inspiring, non-trivial, applications and examples give motivation and ease the learning of abstract concepts. This book is built on several years of experience teaching introductory abstract algebra at Aarhus, where the emphasis on concrete examples has improved student performance significantly.

CONCRETE ABSTRACT ALGEBRA

From Numbers to Gröbner Bases

NIELS LAURITZEN

Department of Mathematical Sciences
University of Aarhus
Denmark

CAMBRIDGE
UNIVERSITY PRESS

CAMBRIDGE UNIVERSITY PRESS
Cambridge, New York, Melbourne, Madrid, Cape Town, Singapore, São Paulo

Cambridge University Press
The Edinburgh Building, Cambridge CB2 8RU, UK

Published in the United States of America by Cambridge University Press, New York

www.cambridge.org
Information on this title: www.cambridge.org/9780521826792

First published 2003
Fourth printing 2007

Printed in the United Kingdom at the University Press, Cambridge

A catalogue record for this book is available from the British Library

Library of Congress Cataloguing in Publication data

Lauritzen, Niels, 1964–
Concrete abstract algebra: from numbers to Gröbner bases / Niels Lauritzen.
p. cm.
Includes bibliographical references and index.
ISBN 0 521 82679 9 (hardback) – ISBN 0 521 53410 0 (paperback)
1. Algebra, abstract. 1. Title
QA162.L43 2003
512′.02–dc21 2003051248

ISBN 978-0-521-82679-9 hardback
ISBN 978-0-521-53410-9 paperback

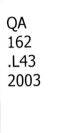
For Helle and William

Contents

Preface

Imagine that you have a very persistent piano teacher insisting that you study notes and practice scales for three years before you are allowed to listen to or play any real music. How is that going to affect your level of inspiration? Are you going to attend every lesson with passion or practice absolutely ignited with energy? Abstract algebra is like piano playing. You can kill your inspiration and motivation spending years on formalism before seeing the beauty of the subject. This book is written with the intent that every chapter should contain some real music, matters which involve practice of the notes and scales in a surprising and unexpected way. It is an attempt to include a lot of non-trivial and fun topics in an introductory abstract algebra course. Having inspiring goals makes the learning easier. The topics covered in this book are numbers, groups, rings, polynomials and Gröbner bases.

Knowledge of linear algebra and complex numbers is assumed in some examples. However, most of the text is accessible with only basic mathematical topics such as sets, maps, elementary logic and proofs.

Gröbner bases are usually not treated at an undergraduate level. My feeling four years ago when including this topic in the syllabus at Aarhus was one of hesitation. I was afraid that the material would be too advanced for the students. It turned out that the students liked the concrete nature of the material and enjoyed the non-trivial computations with polynomials. They found it easier than the traditional topics of groups and rings.

Unlike most treatments on Gröbner bases, I have not included any implementations of algorithms in a pseudo-language. My personal experience is that it disturbs the flow of the mathematics when teaching the basic ideas of the algorithms. Once the mathematical concepts and a few examples are understood, it is easy to extract the algorithms for implementation on a computer. In fact students are very much encouraged to experiment using a computer algebra system especially when learning about numbers and Gröbner bases.

Chapter 1 is on numbers. It is mostly based on the RSA cryptosystem and the mystery that it seems much easier to multiply numbers than to factor them. The 617-digit number on the cover of this book is a product of two prime numbers. If you can find them you should write to RSA Labs and claim the $200, 000 prize. Going through the first chapter you will learn basic number theory: division with remainder, congruences, the Euclidean algorithm, the Chinese remainder theorem, prime numbers, how prime numbers uncovered the infamous FDIV bug in Intel's Pentium processor, Fermat's little theorem and how it is used to produce 100-digit prime numbers for the modern information age, three modern algorithms for factoring numbers much faster than by trial division, quadratic residues and the quadratic reciprocity theorem (which will be proved in Chapter 4).

The level of abstraction is increased in Chapter 2. Here the mathematical object is a group. A group is defined using a composition on a set and it satisfies three simple rules. This definition has proved extremely important and invaluable to modern algebra. You get a framework for many proofs and concepts from basic number theory. We treat the basics of group theory, the symmetric and alternating groups, how to solve the 15-puzzle using groups, actions of groups, counting and the Sylow theorems.

In Chapter 3 we treat rings. A ring is an abelian group with multiplication as an added composition. We touch briefly on non-commutative rings, with the quaternions as an example. We then move on to commutative rings, Freshman's Dream, fields, domains, principal ideal domains, Euclidean domains and unique factorization domains. The Fermat two-square theorem (every prime number leaving a remainder of 1 when divided by 4 can be written as a sum of two unique squares (e. g. $13 = 3^2 + 2^2$)) is a prime example in this chapter. You will see the infinitude of prime numbers leaving a remainder of 1 when divided by 4, further use of quadratic residues and an effective algorithm for computing the two squares in the two-square theorem.

Polynomials form a central topic. In Chapter 4 we treat polynomials in one variable. Here the highlights are: cyclotomic polynomials, a proof of the law of quadratic reciprocity using only basic properties of rings of polynomials, how to use floating point arithmetic to compute the order of specific elements in a well known cyclic group, the ElGamal cryptosystem, the infinitude of prime numbers congruent to 1 modulo a natural number > 1 and the existence and uniqueness of finite fields, along with algorithms for factoring polynomials over finite fields.

In Chapter 5 polynomials in several variables and Gröbner bases are treated. Gröbner bases form an exciting and relatively new branch of algebra. They are very concrete and computational. The distance from understanding the abstract concepts involved to computing with them is small. They provide a framework for solving non-linear equations (used in most computer algebra systems) with applications in many areas inside and outside algebra. In Chapter 5 you will see term orders, the fundamental Dickson's lemma, the division algorithm for polynomials in several variables, the existence of Gröbner bases, Hilbert's basis theorem, Buchberger's S-criterion and algorithm, how to write $X^4 + Y^4$ as a polynomial in $X + Y$ and XY (like writing $X^2 + Y^2$ as $(X + Y)^2 - 2XY$) using Gröbner bases and how to solve certain non-linear equations in several variables systematically.

A few exercises are marked **HOF**. This indicates that they are "hall of fame" exercises, far beyond what is required in an introductory abstract algebra course. They usually call for an extraordinary amount of ingenuity. A student capable of solving one of these deserves to be inducted into the hall of fame of creative problem solvers. A hall of fame museum can be suitably maintained using a course home page.

Suggestions for teaching a one-semester course

The book contains too much material for a one-semester course in introductory abstract algebra. So, a selection of material must be made. A possible procedure would be to leave out factoring algorithms from Chapter 1, quadratic reciprocity from Chapters 1 and 4 and the Sylow theorems from Chapter 2. This plan would give a one-semester course ending with Gröbner bases; it would cover the usual topics in an introductory course.

Leaving out Gröbner bases completely, Chapters 1 through 4 would form an in-depth traditional introductory abstract algebra course with many examples.

Acknowledgements

I wish to thank all the students of Algebra 1 at the University of Aarhus during the past four years for carefully listening, asking questions, looking puzzled at the right (or wrong) times and for inspiring me to change my exposition several times. I wish in particular to thank R. Villemoes for many valuable comments and for a set of detailed TEX-solutions to the exercises (available through Cambridge University Press).

Many people influenced this book either by discussions and comments or by patiently answering my numerous questions: T. B. Andersen, H. H. Andersen, M. Bökstedt, J. Brandt, A. Buch, A. L. Christophersen, I. Damgaard, R. Faber Larsen, P. de Place Friis, S. Galatius Smith, W. J. Haboush, J. P. Hansen, G. Hellmund, C. U. Jensen, T. H. Lynderup, T. Høholdt, T. Laframboise, M. Skov Madsen, K. Nielsen, U. Raben Pedersen, M. S. Risager, A. Skovborg, H. G. Spalk, J. Tornehave, H. Vosegaard and A. Venkov.

I am particularly indebted to J. C. Jantzen for reading carefully earlier versions of my Algebra 1 notes. His comments were (as always) extremely relevant and helpful. J. F. Thomsen also read earlier versions of the notes and made detailed comments on the 15-puzzle, which led to substantial improvements. H. A. Salomonsen pointed out a substantial simplification that moved the proof of quadratic reciprocity from the context of finite fields to the more student-friendly environment of the basic theory of polynomials. An anonymous referee from the US made meticulous comments and suggestions which greatly facilitated the process of turning my incomplete notes into the present book. J. Walthoe at Cambridge University Press has been extremely helpful making several insightful suggestions.

This book is for Helle and William. They have unselfishly fueled my writing with their love.

1 Numbers

This chapter serves as an introduction to the modern theory of algebra through the natural numbers $0, 1, 2, \ldots$. The list of natural numbers never ends and most of them are far beyond everyday use. Gigantic numbers of more than 100 digits are used to protect information transmitted over the internet.

Suppose Alice has to send a message to Bob over the internet and it must be kept secret. Alice and Bob live far apart and many intermediate computers will see the message on its way. Alice will have to scramble (encrypt) the message and send it, but at the same time Bob will have to know how to unscramble (decrypt) it. How does Alice get this information through to him? She could call and tell him. But then again someone could be listening in on their phone call. Is there a way out of this problem?

The answer is an amazing "yes" and it builds on a current paradox of mathematics: the existence of so-called one-way functions $f(X)$. These are functions easy to compute given the input X. Once they are computed and only $f(X)$ is known, it appears to be exceedingly difficult to recover X unless some secret information is known.

Here is an example of a one-way function. Fix a natural number N and let $f(X) = [X^3]$, where $[Y]$ denotes the remainder of Y after division by N. This is a function $f : M \to M$, where $M = \{0, 1, 2, \ldots, N - 1\}$. When $N = 15$, f can be tabulated as

X	0	1	2	3	4	5	6	7	8	9	10	11	12	13	14
$f(X)$	0	1	8	12	4	5	6	13	2	9	10	11	3	7	14

Of course we can easily find X given $f(X)$ by using the above table. But in general, as N grows the difficulty of finding X given $f(X)$ seems insurmountable unless you know some secret information. In the above example the secret information is that $f(f(X)) = X$ (you can see this using the table). In a sense we are raising a number to the third power and then scrambling things up by

1

taking the remainder. So far nobody has found effective methods for finding cube roots in this setting. In the above example Alice sends the encrypted message $f(X)$ to Bob and Bob decrypts it using f. This is the basic principle behind the RSA cryptosystem [22], which was the first cryptosystem based on the groundbreaking idea [8] of using one-way functions (with a trapdoor).

On a more detailed level Bob computes two gigantic prime numbers (usually 100 digits or more) p and q and forms $N = pq$. He then uses p and q to compute a number e (for encryption) and a number d (for decryption). He makes the numbers N and e public so that people wishing to write secret messages to him can use the function $f(X) = [X^e]$ for encryption, where $[Y]$ denotes the remainder of Y after division by N. He keeps the function $g(X) = [X^d]$ secret (the point being that $g(f(X)) = X$). In the example above we have $p = 3, q = 5, N = 15, e = 3, d = 3$. One way of systematically finding the secret decryption function g in the RSA system is to find the prime factors p and q of N (N being available to the general public). The straightforward method of trial division (dividing with successive primes $2, 3, 5, \ldots$) is much too slow. Mathematicians have tried at least since Gauss's time (1777–1855) to find faster methods for factoring numbers. In fact Gauss writes in ([11], Art. 329)

> The problem of distinguishing prime numbers from composite numbers and of resolving the latter into their prime factors is known to be one of the most important and useful in arithmetic. It has engaged the industry and wisdom of ancient and modern geometers to such an extent that it would be superfluous to discuss the problem at length. Nevertheless we must confess that all methods that have been proposed thus far are either restricted to very special cases or are so laborious and prolix that even for numbers that do not exceed the limits of tables constructed by estimable men, i.e., for numbers that do not yield to artificial methods, they try the patience of even the practiced calculator. And these methods do not apply at all to larger numbers.

RSA Labs has put forward several factoring challenges. The hardest unsolved challenge is called RSA-2048. This is the 2048-bit number (617 digits) N on the cover of this book. It is known to be the product of two prime numbers p and q. A computer was instructed to forget p and q after forming $N = pq$. Given two candidates p' and q', it is easy to multiply them to see if their product equals N. This can be done in a small fraction of a second on any modern computer. Nevertheless, finding p and q knowing only N seems to be a painstakingly slow process not within the limits of modern computers and algorithms. If you can find p and q you will be able to claim the $200 000 prize by submitting your factorization via http://www.rsasecurity.com/go/factorization.html. Alternatively, you could settle for the less ambitious RSA factoring challenges presented at http://www.rsasecurity.com/rsalabs/challenges/factoring/numbers.html. It has not been proved mathematically that factoring a number is a difficult problem in a precise sense, so a fast algorithm may exist waiting to be discovered. In a sense this would disrupt the pillars of the modern information age. The algebraic reasoning behind the RSA cryptosystem is founded on basic results (more than 300 years old) about the natural numbers.

1.1 The natural numbers and the integers

The natural numbers $1, 2, 3, \ldots$ were handed over to mankind by God (in the words of Kronecker (1823–91)). Mankind later added the important natural number 0. We will reserve the symbol $\mathbb{N}$ for the natural numbers $\{0, 1, 2, 3, \ldots\}$. The need for negative numbers leads us to introduce the set of integers $\mathbb{Z} = \{\ldots, -2, -1, 0, 1, 2, \ldots\}$ containing the natural numbers $\mathbb{N}$. We have deliberately cut through the red tape of formally defining $\mathbb{N}$ and $\mathbb{Z}$ here. We will also take the addition (and subtraction) and multiplication of integers for granted. This will be the starting point of our study of numbers.

1.1.1 Well ordering and mathematical induction

For $X, Y \in \mathbb{Z}$ we define $X \leq Y$ if $Y - X \in \mathbb{N}$ and $X < Y$ if $X \neq Y$ and $X \leq Y$. This leads to the usual way of ordering the integers,

$$\cdots < -3 < -2 < -1 < 0 < 1 < 2 < 3 < \cdots .$$

An element s in a subset $S \subseteq \mathbb{Z}$ is said to be a first element in S if $s \leq x$ for every $x \in S$. There are many subsets of $\mathbb{Z}$ that do not have a first element. If a subset of $\mathbb{Z}$ has a first element then the latter has to be unique (see Exercise 1.1 at the end of the chapter). The basic axiom for starting our investigation of numbers says that *every non-empty subset of $\mathbb{N}$ has a first element.* We also say that the set of natural numbers is *well ordered.*

The property that $\mathbb{N}$ is well ordered is equivalent to mathematical induction. Recall that mathematical induction says that if we are given statements $P(n)$ for every integer $n \geq 1$ such that

(i) $P(1)$ is true and
(ii) $P(n)$ is true implies that $P(n + 1)$ is true

then $P(n)$ is true for every $n \geq 1$.

Example 1.1.1 Let us prove the formula

$$1 + 2 + \cdots + n = \frac{n(n+1)}{2} \tag{1.1}$$

for $n \in \mathbb{N}$ using mathematical induction. This means that we consider (1.1) as a statement $P(n)$. Clearly $P(1)$ is true, since $1 \cdot (1 + 1) = 2$. Suppose now that $P(n)$ is true. Then

$$1 + 2 + \cdots + n + (n+1) = \frac{n(n+1)}{2} + (n+1).$$

The right hand side can be rewritten as

$$\frac{n(n+1)}{2} + (n+1) = \frac{n(n+1) + 2(n+1)}{2}$$
$$= \frac{(n+1)(n+2)}{2}.$$

This is the formula for $n + 1$. So we have proved that $P(n)$ implies $P(n+1)$. By mathematical induction we have proved $P(n)$ for every $n \geq 1$.

Of course, having the formal machinery for constructing a proof like this does not necessarily provide the beauty of a really ingenious mathematical argument. When Gauss was in school (at the age of seven) his mathematics teacher asked the class to sum up all numbers from 1 to 100. The students worked furiously with their small slates. Gauss was the first to give his slate with the number 5050 to the teacher. The teacher replied "Oh, I see, you probably knew the answer." "No, no! I just realized that

$$1 + 100 = 101,$$
$$2 + 99 = 101,$$
$$3 + 98 = 101,$$
$$\vdots$$
$$100 + 1 = 101.$$

Therefore $1 + 2 + \cdots + 100 = (100 \cdot 101)/2 = 5050$," Gauss replied.

1.2 Division with remainder

Suppose that you mark all multiples of 3 on the axis of the integers:

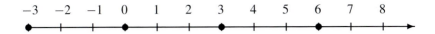

An integer is uniquely given by the closest multiple of 3 to its left and the remainder you have to walk to the right. Examples are $5 = 3 + 2 = 1 \cdot 3 + 2, 7 = 6 + 1 = 2 \cdot 3 + 1, -2 = -3 + 1 = -1 \cdot 3 + 1$ and $6 = 6 + 0 = 2 \cdot 3 + 0$. Division with remainder is the generalization of this simple fact.

Theorem 1.2.1 *Let $d \in \mathbb{Z}$, where $d > 0$. For every $x \in \mathbb{Z}$ there is a unique remainder $r \in \mathbb{N}$ such that*

$$x = qd + r,$$

where $q \in \mathbb{Z}$ and $0 \le r < d$.

Proof. To prove the uniqueness of r assume that $x = q_1 d + r_1$ and $n = q_2 d + r_2$, where $q_1, q_2, r_1, r_2 \in \mathbb{Z}$ and $0 \le r_1, r_2 < d$. Then

$$(q_1 - q_2)d = r_2 - r_1.$$

If $r_1 \ne r_2$ we may assume that $r_2 > r_1$. This implies that $r_2 - r_1 = md$, where $m \ge 1$. But this contradicts the fact that $r_2 - r_1 \le r_2 < d$. To prove the existence of r, let $M = \{x - qd \mid q \in \mathbb{Z}\}$. Then $M \cap \mathbb{N} \ne \emptyset$ (see Exercise 1.2) and we let r be the first element in the subset $M \cap \mathbb{N}$ of $\mathbb{N}$. Now $r = x - qd$ for some q and we claim that $0 \le r < d$. If $r \ge d$ then $r > r - d \ge 0$ and $r - d = x - (q + 1)d \in M \cap \mathbb{N}$. This contradicts that r is the first element in $M \cap \mathbb{N}$. $\square$

Definition 1.2.2 Suppose that $a = bc$ where $a, b, c \in \mathbb{Z}$. Then we say that c is a *divisor* of a (*it divides a*). We write this as $c \mid a$.

Notice that 1 and -1 divide every integer and that 0 only divides 0.

Definition 1.2.3 If $x, d \in \mathbb{Z}$, where $d > 0$, we let $[x]_d$ denote the unique remainder r in Theorem 1.2.1. Sometimes we use the notation $[x]$ when it is clear which d we are using.

1.3 Congruences

Gauss published his monumental work [11] on numbers when he was 24 years old. He had begun his deep studies in the theory of numbers at age 18. At the start of [11] he introduced the theory of congruences, which turned out to be of fundamental importance. Congruences form an elegant way of organizing the integers according to their remainders with respect to a fixed number.

Definition 1.3.1 Let $a, b, c \in \mathbb{Z}$. Then a and b are called *congruent modulo* c if c divides $b - a$. This is denoted

$$a \equiv b \,(\mathrm{mod}\ c).$$

This may seem strange at first, but using remainders the definition (for $c > 0$) just states that a and b are congruent modulo c if and only if a and b have the same remainder when divided by c. This is the content of the following:

Proposition 1.3.2 *Let $c \in \mathbb{Z}$, where $c > 0$. Then*

(i) $a \equiv [a]_c \,(\mathrm{mod}\ c)$,
(ii) $a \equiv b \,(\mathrm{mod}\ c)$ *if and only if* $[a]_c = [b]_c$,

for $a, b \in \mathbb{Z}$.

Proof. We may write $a = qc + [a]_c$ for some $q \in \mathbb{Z}$, by Theorem 1.2.1. Therefore $c \mid a - [a]_c = qc$. This proves (i). Now write $b = q'c + [b]_c$ for some $q' \in \mathbb{Z}$. Then $a - b = (q - q')c + [a]_c - [b]_c$. Therefore $c \mid a - b$ if and only if $c \mid [a]_c - [b]_c$. But $c \mid [a]_c - [b]_c$ if and only if $[a]_c = [b]_c$, since $0 \le [a]_c, [b]_c < c$. This proves (ii). $\qquad\square$

Example 1.3.3 The integers 24 and 14 can be written $24 = 4 \cdot 5 + 4$ and $14 = 2 \cdot 5 + 4$. So $[24]_5 = [14]_5 = 4$. This means that $24 \equiv 14$ (mod 5). Of course this could just as easily have been observed from the fact that $5 \mid 24 - 14$.

Proposition 1.3.4 *Suppose that $x_1 \equiv x_2 \,(\mathrm{mod}\ d)$ and $y_1 \equiv y_2 \,(\mathrm{mod}\ d)$. Then*

(i) $x_1 + y_1 \equiv x_2 + y_2 \,(\mathrm{mod}\ d)$,
(ii) $x_1 y_1 \equiv x_2 y_2 \,(\mathrm{mod}\ d)$

for $x_1, x_2, y_1, y_2, d \in \mathbb{Z}$.

Proof. If d divides $x_1 - x_2$ and $y_1 - y_2$ then it also divides $x_1 - x_2 + y_1 - y_2 = x_1 + y_1 - (x_2 + y_2)$. This proves (i). Rearranging, we also get that d divides $x_1 y_1 - x_2 y_2 = x_1(y_1 - y_2) + y_2(x_1 - x_2)$. This proves (ii). $\qquad\square$

Proposition 1.3.4 may look innocuous at first. It is surprisingly useful. For one thing, when you combine it with Proposition 1.3.2, you get (see Exercise 1.3)

$$[xy] = [[x][y]]. \tag{1.2}$$

Using (1.2) you can tell in a flash that the remainder of 13^{2003} divided by 4 has to be 1 (how?). Take a look at the following example.

1.3.1 Repeated squaring – an example

How does one find the remainder of 12^{11} divided by 21 efficiently? This problem confronts a sender of a secret message in the RSA cryptosystem, where the encryption exponent is the number $e = 11$ and the possible messages are the natural numbers less than $N = 21$. As you may have guessed the trick is to avoid computing the integer 12^{11}, divide by 21 and find the remainder. First we write 11 in the binary expansion (11 can be expressed as 1011 in the binary positional system) as

$$2^3 + 2 + 1.$$

Then using (1.2) twice we see that

$$[12^{11}] = \left[12^{2^3}\, 12^2\, 12^1\right] = \left[\left[12^{2^3}\right][12^2][12^1]\right].$$

Again using (1.2) we build a table of remainders for use in the calculation

$$
\begin{aligned}
[12^1] &= 12, \\
[12^2] &= 18, \\
\left[12^{2^2}\right] &= [(12^2)^2] = [[12^2][12^2]] = [18 \cdot 18] = 9, \\
\left[12^{2^3}\right] &= [(12^{2^2})^2] = \left[\left[12^{2^2}\right]\left[12^{2^2}\right]\right] = [9 \cdot 9] = 18.
\end{aligned}
$$

Picking out the relevant numbers we get

$$
\begin{aligned}
[12^{11}] &= [[18 \cdot 18] \cdot 12] \\
&= [9 \cdot 12] \\
&= 3.
\end{aligned}
$$

We have reduced the horrendous procedure of computing the remainder of $12^{11} = 743008370688$ divided by 21 to computing the remainders of numbers less than $21^2 = 441$. The algorithm above is called *repeated squaring*, because we constantly use the following consequence of (1.2):

$$\left[a^{2^n}\right] = \left[\left(a^{2^{n-1}}\right)^2\right] = \left[\left[a^{2^{n-1}}\right]\left[a^{2^{n-1}}\right]\right],$$

for $a, n \in \mathbb{Z}$ where $n \geq 0$ (recall that $(a^b)^c = a^{bc}$, where $a, b, c \in \mathbb{Z}$ with $b, c \geq 0$).

1.4 Greatest common divisor

Let

$$\mathrm{div}(n) = \{d \in \mathbb{N} \mid d \mid n\}$$

denote the set of natural divisors in $n \in \mathbb{Z}$. Notice that $\mathrm{div}(0) = \mathbb{N}$ and $\mathrm{div}(n) = \mathrm{div}(-n)$ for every $n \in \mathbb{Z}$.

Example 1.4.1   Let us list a few examples:

(i) $\mathrm{div}(18) = \{1, 2, 3, 6, 9, 18\}$,
(ii) $\mathrm{div}(24) = \{1, 2, 3, 4, 6, 8, 12, 24\}$,
(iii) $\mathrm{div}(36) = \{1, 2, 3, 4, 6, 9, 12, 18, 36\}$.

From this example we have

$$\mathrm{div}(24) \cap \mathrm{div}(18) = \{1, 2, 3, 6\} = \mathrm{div}(6)$$
$$\mathrm{div}(24) \cap \mathrm{div}(36) = \{1, 2, 3, 4, 6, 12\} = \mathrm{div}(12).$$

This indicates a striking fact. Given two integers m, n it seems that the common divisors $\mathrm{div}(m) \cap \mathrm{div}(n)$ of m and n are exactly the divisors $\mathrm{div}(d)$ of some third number. This is not a coincidence. It was discovered by the Greek mathematician Euclid of Alexandria (325–265BC) and is contained in book seven of his masterpiece, the *Elements*.

Lemma 1.4.2 (Euclid)   *Let $m, n \in \mathbb{Z}$. There exists a unique natural number $d \in \mathbb{N}$ such that*

$$\mathrm{div}(m) \cap \mathrm{div}(n) = \mathrm{div}(d).$$

Proof.   The uniqueness follows from the fact that $\mathrm{div}(d_1) = \mathrm{div}(d_2)$ if and only if $d_1 = d_2$ assuming that $d_1, d_2 \in \mathbb{N}$. When proving the existence of d we may assume that $m, n \in \mathbb{N}$, since $\mathrm{div}(x) = \mathrm{div}(-x)$ for $x \in \mathbb{Z}$. We proceed using induction on $\min(m, n)$, where $\min(m, n) = m$ if $m \leq n$ and $\min(m, n) = n$ if $m > n$. If $\min(m, n) = 0$ we may assume that $n = 0$. Therefore $\mathrm{div}(m) \cap \mathrm{div}(n) = \mathrm{div}(m)$. This settles the initial step $\min(m, n) = 0$ of the induction.

Now assume that we have proved $\mathrm{div}(m) \cap \mathrm{div}(n) = \mathrm{div}(d)$ for every $m, n \in$ $\mathbb{N}$ with $\min(m, n) < N$, where $N > 0$. Suppose for the induction step that we are given $m, n \in \mathbb{N}$ with $\min(m, n) = N$ and that $m \geq n = N$. Then we may write $m = qn + r$, where $0 \leq r < n$ by Theorem 1.2.1. But (this is the clever step)

$$\mathrm{div}(m) \cap \mathrm{div}(n) = \mathrm{div}(m - qn) \cap \mathrm{div}(n) = \mathrm{div}(r) \cap \mathrm{div}(n),$$

since a number divides m and n if and only if it divides $m - qn$ and n. By induction we know that $\mathrm{div}(r) \cap \mathrm{div}(n) = \mathrm{div}(d)$ for some $d \in \mathbb{N}$, since $\min(r, n) = r < n = N$. This completes the proof. $\square$

Definition 1.4.3   The unique number $d \in \mathbb{N}$ satisfying $\mathrm{div}(d) = \mathrm{div}(m) \cap$ $\mathrm{div}(n)$ is called the *greatest common divisor* of m and n. It is denoted $\gcd(m, n)$.

If one of m and n is non-zero there is a finite number of common natural divisors. The greatest common divisor is really the greatest among these with respect to the usual ordering of $\mathbb{Z}$ (see Exercise 1.9). Notice that $\gcd(0, 0) = 0$.

1.5   The Euclidean algorithm

As already hinted in the inductive proof of Lemma 1.4.2, there is an algorithm for finding the greatest common divisor. The inductive step in the proof of Lemma 1.4.2 can be found in Euclid's *Elements* (around 300 BC) even though Euclid did not have the concept of induction and the rigor of a modern mathematical proof. The idea behind the modern version of Euclid's algorithm is the same.

Proposition 1.5.1   *Let $m, n \in \mathbb{Z}$. Then*

(i) $\gcd(m, 0) = m$ *if $m \in \mathbb{N}$.*
(ii) $\gcd(m, n) = \gcd(m - qn, n)$ *for every $q \in \mathbb{Z}$.*

Proof.   Since $\mathrm{div}(0) = \mathbb{N}$, (i) follows. We get (ii) from the fact that

$$\mathrm{div}(m) \cap \mathrm{div}(n) = \mathrm{div}(m - qn) \cap \mathrm{div}(n).$$

This is a way of saying that a natural number d divides m and n if and only if d divides $m - qn$ and n, so that $\gcd(m, n) = \gcd(m - qn, n)$. $\square$

Suppose that we wish to find the greatest common divisor of $m, n \in \mathbb{Z}$. We may assume that $m \geq n \geq 0$. If $n = 0$, we are done since $\gcd(m, 0) = m$ by Proposition 1.5.1(i). Assume that $n > 0$. The basic observation is that if we divide m by n and write $m = qn + r$ according to Theorem 1.2.1, then

$$\gcd(m, n) = \gcd(r, n) = \gcd(n, r)$$

and $n > r$. This follows from Proposition 1.5.1(ii). An example shows how this works.

Example 1.5.2   Let $m = 34$ and $n = 13$. Then

$$\gcd(34, 13) = \gcd(13, 8) = \gcd(8, 5)$$
$$= \gcd(5, 3) = \gcd(3, 2) = \gcd(2, 1)$$
$$= \gcd(1, 0) = 1.$$

This can also be illustrated as a sequence of divisions with remainders:

$$34 = 2 \cdot 13 + 8,$$
$$13 = 1 \cdot 8 + 5,$$
$$8 = 1 \cdot 5 + 3,$$
$$5 = 1 \cdot 3 + 2,$$
$$3 = 1 \cdot 2 + 1,$$
$$2 = 2 \cdot 1 + 0.$$

Now return to the general case $m \geq n \geq 0$. Put $r_{-1} = m$ and $r_0 = n$. If $r_0 = 0$ then $\gcd(r_{-1}, r_0) = r_{-1}$. Otherwise define r_1 to be the remainder of r_{-1} divided by r_0, so that $r_1 = r_{-1} - q_1 r_0$ for some integer q_1. Then we have

$$\gcd(r_{-1}, r_0) = \gcd(r_0, r_1)$$

and $r_{-1} > r_0 > r_1$. Proceeding in this way (if $r_1 \neq 0$) we let $r_2 = r_0 - q_2 r_1$ be the remainder of r_0 divided by r_1. Again we have

$$\gcd(r_0, r_1) = \gcd(r_1, r_2)$$

and $r_{-1} > r_0 > r_1 > r_2$. Eventually we are forced to the situation $r_N = 0$, for some step $N > 0$. This means that $\gcd(m, n) = \gcd(r_{N-1}, 0) = r_{N-1}$. The point is that the Euclidean algorithm gives rise to a strictly decreasing sequence of natural numbers $r_{-1} > r_0 > r_1 > \cdots$. If we consider the subset $R = \{r_{-1}, r_0, r_1, \dots\}$ as a subset of $\mathbb{N}$, it has a first element $r_N \in R$ since $\mathbb{N}$ is well ordered. If $r_N \neq 0$ we may continue division with remainder and get

$r_N > r_{N+1} \geq 0$, contradicting the fact that r_N is the first element in R. Therefore $r_N = 0$ and the Euclidean algorithm terminates in a finite number of steps.

A very important fact is hidden in the Euclidean algorithm: the greatest common divisor $\gcd(m, n)$ can be written as a $\mathbb{Z}$-linear combination of m and n. There exist integers λ and μ such that

$$\lambda m + \mu n = \gcd(m, n).$$

Let us go through the steps in the Euclidean algorithm once more and make a few adjustments.

Example 1.5.3   We know that $\gcd(34, 13) = 1$. The claim above says that one can find integers x and y such that $34x + 13y = 1$. This is not obvious. We need an algorithm for computing x and y. The trick is to adjust x and y for each remainder in the steps of the Euclidean algorithm:

$$34 = 1 \cdot 34 + 0 \cdot 13,$$
$$13 = 0 \cdot 34 + 1 \cdot 13,$$
$$8 = 34 - 2 \cdot 13 = (1 \cdot 34 + 0 \cdot 13) - 2 \cdot (0 \cdot 34 + 1 \cdot 13),$$
$$= 1 \cdot 34 - 2 \cdot 13,$$
$$5 = 13 - 8 = (0 \cdot 34 + 1 \cdot 13) - (1 \cdot 34 - 2 \cdot 13)$$
$$= -1 \cdot 34 + 3 \cdot 13,$$
$$3 = 8 - 5 = (1 \cdot 34 - 2 \cdot 13) - (-1 \cdot 34 + 3 \cdot 13)$$
$$= 2 \cdot 34 - 5 \cdot 13,$$
$$2 = 5 - 3 = (-1 \cdot 34 + 3 \cdot 13) - (2 \cdot 34 - 5 \cdot 13)$$
$$= -3 \cdot 34 + 8 \cdot 13,$$
$$1 = 3 - 2 = (2 \cdot 34 - 5 \cdot 13) - (-3 \cdot 34 + 8 \cdot 13)$$
$$= 5 \cdot 34 - 13 \cdot 13.$$

Attaching these small updates to the Euclidean algorithm we have produced the identity

$$5 \cdot 34 - 13 \cdot 13 = 1,$$

which would have been hard to guess initially.

Definition 1.5.4   The Euclidean algorithm with the above attachment for computing x and y is called the *extended Euclidean algorithm*.

Let us be a little more formal in the description of the extended Euclidean algorithm. Define at each step of the algorithm integers a_i and b_i with the property that $a_i m + b_i n = r_i$. One can start by putting $a_{-1} = 1$, $b_{-1} = 0$ and $a_0 = 0$, $b_0 = 1$. The first step of the algorithm is $r_1 = r_{-1} - q_1 r_0$. The definition of a_1 and b_1 leaves no choice: $a_1 = a_{-1} - q_1 a_0$, $b_1 = b_{-1} - q_1 b_0$. The ith step proceeds similarly, as $r_i = r_{i-2} - q_i r_{i-1}$ (where $r_{i-2} = q_i r_{i-1} + r_i$ according to Theorem 1.2.1). This means that for $i \geq 1$ we put

$$a_i = a_{i-2} - q_i a_{i-1},$$
$$b_i = b_{i-2} - q_i b_{i-1}.$$

Assuming that $a_{i-1} m + b_{i-1} n = r_{i-1}$ and $a_{i-2} m + b_{i-2} n = r_{i-2}$ this ensures that

$$a_i m + b_i n = (a_{i-2} - q_i a_{i-1})m + (b_{i-2} - q_i b_{i-1})n$$
$$= a_{i-2} m + b_{i-2} n - q_i(a_{i-1} m + b_{i-1} n)$$
$$= r_{i-2} - q_i r_{i-1}$$
$$= r_i.$$

The extended Euclidean algorithm is conveniently carried out using the table in the example below.

Example 1.5.5 The greatest common divisor of 13 and 8 is 1. Illustrated in the table below is the extended Euclidean algorithm, giving $-3 \cdot 13 + 5 \cdot 8 = 1$.

i	-1	0	1	2	3	4
r_i	13	8	5	3	2	1
q_i			1	1	1	1
a_i	1	0	1	-1	2	-3
b_i	0	1	-1	2	-3	5

Remark 1.5.6 Which numbers less than a given number result in the maximum number of steps in the Euclidean algorithm? To answer this question we need to define the Fibonacci numbers F_n. They are given by $F_0 = 1$, $F_1 = 1$ and $F_n = F_{n-1} + F_{n-2}$ for $n \geq 2$. The first few Fibonacci numbers are $1, 1, 2, 3, 5, 8, 13, 21, \ldots$. These numbers have a surprising relation ([16], subsection 4.5.3) to the complexity of the Euclidean algorithm: if $u > v > 0$ are integers, and u is the smallest number such that the Euclidean algorithm for u and v needs exactly n steps, then $u = F_{n+1}$ and $v = F_n$.

This result dates back to 1845 and is due to Lamé. Knuth [16] writes that it has the historical claim of being the first practical application of Fibonacci numbers.

Let us reiterate the very important fact contained in the extended Euclidean algorithm. It is the basis of almost all the results in the rest of this chapter.

Lemma 1.5.7   *Let $m, n \in \mathbb{Z}$. Then there are integers $\lambda, \mu \in \mathbb{Z}$ such that*

$$\lambda m + \mu n = \gcd(m, n).$$

Proof.   Let $d = \gcd(m, n)$. The extended Euclidean algorithm gives this result if $m, n \in \mathbb{N}$. In this case we can find $\lambda, \mu \in \mathbb{Z}$ such that $\lambda m + \mu n = d$. Notice that $(-\lambda)(-m) + \mu n = \lambda m + (-\mu)(-n) = (-\lambda)(-m) + (-\mu)(-n) = d$. So it is easy to get the result for $m, n \in \mathbb{Z}$. ☐

Definition 1.5.8   Two integers $a, b \in \mathbb{Z}$ are called *relatively prime* if

$$\gcd(a, b) = 1.$$

Remark 1.5.9   Notice that if there are $\lambda, \mu \in \mathbb{Z}$ such that $\lambda a + \mu b = 1$ then a and b are relatively prime (see Exercise 1.14).

Corollary 1.5.10   *Suppose that $a \mid bc$, where $a, b, c \in \mathbb{Z}$ and a and b are relatively prime. Then $a \mid c$.*

Proof.   According to Lemma 1.5.7, we may find $\lambda, \mu \in \mathbb{Z}$ such that $\lambda a + \mu b = 1$. Multiply this equation by c and get $\lambda ac + \mu bc = c$. Now a divides the left hand side, since a divides bc. Therefore a divides c. ☐

Corollary 1.5.11   *Let $a, b, c \in \mathbb{Z}$.*

 (i) *If a and b are relatively prime, $a \mid c$ and $b \mid c$ then $ab \mid c$.*
 (ii) *If a and b are relatively prime and a and c are relatively prime then a and bc are relatively prime.*

Proof.   Since $\gcd(a, b) = 1$ we get $\lambda a + \mu b = 1$ for suitable $\lambda, \mu \in \mathbb{Z}$ by Lemma 1.5.7. Both a and b divide c, so we may write $c = ax = by$ for suitable $x, y \in \mathbb{Z}$. Then

$$c = c(\lambda a + \mu b) = c\lambda a + c\mu b = by\lambda a + ax\mu b = ab(y\lambda + x\mu).$$

This proves (i). To prove (ii), we again use Lemma 1.5.7. This time we get two identities $\lambda a + \mu b = 1$ and $\lambda_1 a + \mu_1 c = 1$ for suitable $\lambda, \mu, \lambda_1, \mu_1 \in \mathbb{Z}$.

Multiplying these we get

$$(\lambda\lambda_1 a + \lambda\mu_1 c + \lambda_1\mu b)a + \mu\mu_1 bc = 1.$$

This shows that $\gcd(a, bc) = 1$, so that a and bc must be relatively prime. □

In trying to grasp statements as Corollaries 1.5.10 and 1.5.11, it often pays to play with small numbers to find counter-examples, such as the simple fact that $4 \mid 2 \cdot 2$ but $4 \nmid 2$. Also, $6 \mid 12$ and $3 \mid 12$ but $6 \cdot 3 = 18 \nmid 12$.

1.6 The Chinese remainder theorem

Think of a natural number x less than 30. Let a, b, c respectively denote the rows (numbered upwards from zero) in the three tables below, in which the number is located.

0	2	4	6	8	10	12	14	16	18	20	22	24	26	28
1	3	5	7	9	11	13	15	17	19	21	23	25	27	29

0	3	6	9	12	15	18	21	24	27
1	4	7	10	13	16	19	22	25	28
2	5	8	11	14	17	20	23	26	29

0	5	10	15	20	25
1	6	11	16	21	26
2	7	12	17	22	27
3	8	13	18	23	28
4	9	14	19	24	29

For example, if $x = 14$ then $a = 0, b = 2$ and $c = 4$. The real surprise is that one needs only to know these three row numbers in order to determine the original number. This is called the 30-riddle. It has impressed many souls unspoiled by abstract algebra and number theory. The most hard-core algebraists will say it is trivial, referring to the fundamental isomorphism $\mathbb{Z}/30 \cong \mathbb{Z}/2 \times \mathbb{Z}/3 \times \mathbb{Z}/5$. Let us expand a little on this theme.

Definition 1.6.1 Define

$$\mathbb{Z}/N = \{X \in \mathbb{N} \mid 0 \le X < N\},$$

for $N \in \mathbb{N}$. If $N = n_1 \cdots n_t \neq 0$ is the product of $n_1, \ldots, n_t \in \mathbb{N}$ we let

$$r : \mathbb{Z}/N \to \mathbb{Z}/n_1 \times \cdots \times \mathbb{Z}/n_t$$

be the map given by $r(X) = ([X]_{n_1}, \ldots, [X]_{n_t})$. We call r the *remainder map*.

Example 1.6.2   Let $N = 2 \cdot 3 \cdot 5 = 30$ and $x = 14$. Then

$$r(x) = (0, 2, 4).$$

This corresponds to the fact that 14 is in row 0 of the first table, row 1 of the second table and row 4 of the third table.

The secret to unlocking the 30-riddle is contained in the following lemma.

Lemma 1.6.3   *Suppose that $N = n_1 \cdots n_t$, where $n_1, \ldots, n_t \in \mathbb{N} \setminus \{0\}$ and $\gcd(n_i, n_j) = 1$ if $i \neq j$. Then the remainder map*

$$r : \mathbb{Z}/N \to \mathbb{Z}/n_1 \times \cdots \times \mathbb{Z}/n_t$$

is bijective.

Proof.   If $r(X) = r(Y)$ then $n_1 \mid X - Y, \ldots, n_t \mid X - Y$ by Proposition 1.3.2(i). Repeated application of Corollary 1.5.11 gives $N = n_1 \cdots n_t \mid X - Y$. Since $0 \leq X, Y < N$, the only way that this is possible is if $X = Y$, so r must be injective. This implies that r is bijective, since it is an injective map between two sets with the same number of elements. $\square$

Lemma 1.6.3 explains the 30-riddle in the sense that a natural number less than 30 is uniquely given by its remainders by division with 2, 3 and 5. The only practical problem is to find a way to compute the inverse map r^{-1}. This is the map you need in order to impress your friends by practicing magic with the 30-riddle. We move on to state and prove the more classical version of Lemma 1.6.3 known as the Chinese remainder theorem (the theorem can be traced back to the Chinese mathematicians Sun-Tsu (around 280–473) and Chin Chiu Shao (1247)). At the end of the proof you will see how to compute the map r^{-1}.

Theorem 1.6.4 (Chinese remainder theorem)   *Suppose that* $N = n_1 \cdots n_t$, *where* $n_1, \ldots, n_t \in \mathbb{Z} \setminus \{0\}$ *and* $\gcd(n_i, n_j) = 1$ *for* $i \neq j$. *Consider the system*

$$X \equiv a_1 \ (\text{mod } n_1),$$
$$X \equiv a_2 \ (\text{mod } n_2),$$
$$\vdots \qquad\qquad\qquad (1.3)$$
$$X \equiv a_t \ (\text{mod } n_t)$$

of congruences for $a_1, \ldots, a_t \in \mathbb{Z}$. *Then*

(i) *(1.3) has a solution* $X \in \mathbb{Z}$.
(ii) *If* $X, Y \in \mathbb{Z}$ *are solutions of (1.3) then* $X \equiv Y \ (\text{mod } N)$. *If* X *is a solution of (1.3) and* $Y \equiv X \ (\text{mod } N)$ *then* Y *is a solution of (1.3)*.

Proof.   We will prove (ii) first. If X, Y are two solutions of (1.3) then $X \equiv a_j$ (mod n_j) and $Y \equiv a_j$ (mod n_j) for $j = 1, \ldots, t$. Therefore $X \equiv Y(\text{mod } n_j)$ (see Exercise 1.11). So $n_j \mid X - Y$, $j = 1, \ldots, t$ and since the n_j are relatively prime, we get (by repeated application of Corollary 1.5.11) that $N = n_1 \cdots n_t$ divides $X - Y$ or $X \equiv Y(\text{mod } N)$. However, if $Y \equiv X \ (\text{mod } N)$ then $Y \equiv X$ (mod n_j) for $j = 1, \ldots, t$. In this case, Y also solves (1.3). This proves (ii).

   The proof of (i) comes from the extended Euclidean algorithm (Lemma 1.5.7) and the fact that n_j and N/n_j are relatively prime (by repeated application of Corollary 1.5.11): we can find integers λ_j, μ_j such that

$$\lambda_1 n_1 + \mu_1 \, N/n_1 = 1,$$
$$\lambda_2 n_2 + \mu_2 \, N/n_2 = 1,$$
$$\vdots$$
$$\lambda_t n_t + \mu_t \, N/n_t = 1.$$

These identities give the useful numbers $A_j = \mu_j(N/n_j)$ for $j = 1, \ldots, t$. Notice that $A_j \equiv 1 \ (\text{mod } n_j)$ and $A_j \equiv 0 \ (\text{mod } n_i)$ if $i \neq j$. We can build a solution from these by putting

$$X = a_1 A_1 + \cdots + a_t A_t.$$

You can check immediately that X solves (1.3). $\square$

   The following example shows how the map r^{-1} is computed using the proof of Theorem 1.6.4(i).

Example 1.6.5 Let us test our knowledge on the 30-riddle itself. Here $n_1 = 2$, $n_2 = 3$ and $n_3 = 5$. The first step is to find $\lambda_i, \mu_i \in \mathbb{Z}$ such that

$$\lambda_1 n_1 + \mu_1 N/n_1 = 2\lambda_1 + 15\mu_1 = 1,$$
$$\lambda_2 n_2 + \mu_2 N/n_2 = 3\lambda_2 + 10\mu_2 = 1,$$
$$\lambda_3 n_3 + \mu_3 N/n_3 = 5\lambda_3 + 6\mu_3 = 1.$$

Here we can take $\lambda_1 = -7$, $\mu_1 = 1$, $\lambda_2 = -3$, $\mu_2 = 1$, $\lambda_3 = -1$, $\mu_3 = 1$. Therefore we get $A_1 = 15$, $A_2 = 10$, $A_3 = 6$ and

$$X = 15a_1 + 10a_2 + 6a_3$$

as a solution to the system

$$X \equiv a_1 \ (\text{mod } 2),$$
$$X \equiv a_2 \ (\text{mod } 3),$$
$$X \equiv a_3 \ (\text{mod } 5)$$

of congruences. By taking the remainder of X after division by 30 we get the number X', $0 \le X' < 30$, solving the 30-riddle. If $a_1 = 0$, $a_2 = 2$, $a_3 = 4$, we get $X = 20 + 24 = 44$. This gives $X' = [X]_{30} = 14$ as expected.

1.7 Euler's theorem

Let

$$(\mathbb{Z}/N)^* = \{X \in \mathbb{Z}/N \mid \gcd(X, N) = 1\}$$

for $N \in \mathbb{N}$ and define the function $\varphi(N) = |(\mathbb{Z}/N)^*|$. This function is the famous Euler φ-function. It counts the numbers relatively prime to and smaller than a given natural number. The beginning of the table of values looks like

n	0	1	2	3	4	5	6	7	8	9	10	11	12	13
$\varphi(n)$	0	1	1	2	2	4	2	6	4	6	4	10	4	12

If you can come up with an effective way of computing φ you will have broken the RSA cryptosystem. The above table was constructed by listing the numbers less than n and counting the ones relatively prime to n. This is a terribly slow way of computing φ. There is a better way, which is still not good enough. It is based on factoring the number n and use of the Chinese remainder theorem. From the table above it is clear that one cannot expect $\varphi(mn) = \varphi(m)\varphi(n)$ for general numbers m and n. Once again the key notion is that of relatively prime numbers.

Proposition 1.7.1   *Let m and n be relatively prime natural numbers. Then*

$$\varphi(mn) = \varphi(m)\varphi(n).$$

Proof.   Put $N = mn$ and let $r : \mathbb{Z}/N \to \mathbb{Z}/m \times \mathbb{Z}/n$ be the remainder map. We know that r is a bijective map by Lemma 1.6.3. If we can prove that

$$r((\mathbb{Z}/N)^*) = (\mathbb{Z}/m)^* \times (\mathbb{Z}/n)^*$$

then we are done, since r then restricts to give a bijective map from $(\mathbb{Z}/N)^*$ to $(\mathbb{Z}/m)^* \times (\mathbb{Z}/n)^*$. Thus we need to prove that $\gcd(X, N) = 1$ if and only if $\gcd([X]_m, m) = 1$ and $\gcd([X]_n, n) = 1$.

Recall that $\gcd(a, c) = \gcd(c, [a]_c)$ for $a, c \in \mathbb{Z}$ with $c > 0$, by Proposition 1.5.1(ii). So $\gcd([X]_m, m) = 1$ and $\gcd([X]_n, n) = 1$ if and only if $\gcd(X, m) = 1$ and $\gcd(X, n) = 1$. It follows by Corollary 1.5.11 that $\gcd(X, m) = 1$ and $\gcd(X, n) = 1$ if and only if $\gcd(X, mn) = 1$. This proves that $\gcd(X, N) = 1$ if and only if $\gcd([X]_m, m) = 1$ and $\gcd([X]_n, n) = 1$.   □

Let us state and prove the main theorem, which is due to Euler (1707–83).

Theorem 1.7.2 (Euler)   *Let $a, n \in \mathbb{Z}$ be relatively prime integers, where $n \in \mathbb{N}$. Then*

$$a^{\varphi(n)} \equiv 1 (\mathrm{mod}\ n).$$

Proof.   First list the $\varphi(n)$ numbers less than and relatively prime to n:

$$0 \le a_1 < a_2 < \cdots < a_{\varphi(n)} < n.$$

As a key point we will prove that

$$\{[aa_1], \ldots, [aa_{\varphi(n)}]\} = \{a_1, \ldots, a_{\varphi(n)}\}, \tag{1.4}$$

where we consider remainders with respect to n. Now, $[aa_i] = [aa_j]$ implies that $aa_i \equiv aa_j \ (\mathrm{mod}\ n)$ by Proposition 1.3.2. Therefore $n \mid a(a_i - a_j)$. Since $\gcd(n, a) = 1$ we have $n \mid a_i - a_j$ by Corollary 1.5.10. This is only possible when $a_i = a_j$ or $i = j$. Thus $[aa_i] \ne [aa_j]$ when $i \ne j$. Notice that $\gcd(n, aa_i) = 1$ by Corollary 1.5.11. This implies that $\gcd(n, aa_i) = \gcd(n, [aa_i]) = 1$ by Proposition 1.5.1(ii). To sum up, we have $\varphi(n)$ different numbers $[aa_1], \ldots, [aa_{\varphi(n)}] \in \mathbb{Z}/n$ all having greatest common divisor 1 with n. The only way this is possible is by having the identity in (1.4). This identity gives

$$[aa_1][aa_2]\cdots[aa_{\varphi(n)}] = a_1 a_2 \cdots a_{\varphi(n)}.$$

Since $aa_i \equiv [aa_i] \,(\text{mod } n)$ by Proposition 1.3.2(i), we get

$$a^{\varphi(n)} a_1 a_2 \cdots a_{\varphi(n)} \equiv a_1 a_2 \cdots a_{\varphi(n)} (\text{mod } n),$$

so that

$$n \mid a_1 \cdots a_{\varphi(n)} \big(a^{\varphi(n)} - 1 \big).$$

By repeated application of Corollary 1.5.11 we get $\gcd(n, a_1 \cdots a_\varphi(n)) = 1$. This shows that $n \mid a^{\varphi(n)} - 1$ by Corollary 1.5.10. Therefore $a^{\varphi(n)} \equiv 1$ (mod n). $\qquad\square$

After having learned a little group theory we will be able to give a really elegant proof of Euler's theorem. This will be a prime example of how things become easier once you find the right (abstract) framework.

1.8 Prime numbers

A prime number is a natural number $p > 1$ that cannot be expressed as a product of natural numbers strictly less than p. In our notation this means that div(p) = $\{1, p\}$. This is a fundamental definition. The natural number 1 is of a different nature, since it divides every integer. It is easy to decide whether a given number x is relatively prime to a prime number p: it happens if and only if $p \nmid x$ (why?). This implies that $\varphi(p) = p - 1$. We will compute φ for powers of a prime number in subsection 1.8.3. The list of prime numbers begins

$$2, 3, 5, 7, 11, 13, 17, 19, 23, \ldots$$

and can be generated by a beautiful classical method known as the sieve of Eratosthenes, as follows. List the natural numbers > 1:

$$2, 3, 4, 5, 6, 7, 8, 9, 10, 11, 12, 13, 14, 15, 16, 17, 18, 19, 20, 21, 22, 23, \ldots$$

Begin by crossing out all the numbers divisible by 2 (except 2). Move on to the next available number, which is not crossed out (3), cross out all numbers divisible by 3 (except 3) and so on. This leads to the sequence

$$2, 3, \times, 5, \times, 7, \times, \times, \times, 11, \times, 13, \times, \times, \times, 17, \times, 19, \times, \times, \times, 23, \ldots,$$

where the numbers left have to be prime numbers. The number

$$2^{32,582,657} - 1,$$

discovered on September 4, 2006 is currently the largest prime number known to man. This is a number with over four million digits. Without the use of a computer, Lucas (1842–91) proved in 1876 that

$$2^{127} - 1 = 170141183460469231731687303715884105727$$

is a prime number. This was referred to as having a huge number of digits, 39 [18], in 1948. A prime number of the form $M_n = 2^n - 1$ is called a Mersenne prime number (named after the French monk Marin Mersenne (1588–1648)). There is hectic activity on the internet searching for new Mersenne prime numbers (this project is called GIMPS — the Great Internet Mersenne Prime Search). Skilled programmers developed the settings for the project, in which you can participate using the idle CPU-seconds on your personal computer. Currently a \$100 000 prize (from the Electronic Frontier Foundation) is offered to the person(s) discovering the first ten-million-digit prime number. Using the URL http://www.mersenne.org/ you may catch up with the current status of GIMPS.

1.8.1 There are infinitely many prime numbers

It is not known whether there are infinitely many Mersenne prime numbers. Euclid proved that there are infinitely many prime numbers. This proof is more than 2000 years old and still breathtaking. First we need a lemma.

Lemma 1.8.1 *Every non-zero natural number n is a product of prime numbers.*

Proof. The natural number 1 is the empty product of prime numbers by definition. We prove the general statement by induction. Assume that every natural number $m < n$ is a product of prime numbers. Then we have to prove that n is a product of prime numbers. If n is a prime number then it is a product of prime numbers (with one factor). If n is not a prime number then

$$n = n_1 n_2$$

where n_1 and n_2 are natural numbers strictly less than n. By induction, n_1 and n_2 are products of prime numbers. Therefore n is a product of prime numbers.   □

Theorem 1.8.2 (Euclid)   *There are infinitely many prime numbers.*

Proof.   Suppose that there are only finitely many prime numbers, listed as

$$p_1, p_2, \ldots, p_n.$$

Now form the integer $N = p_1 \cdots p_n + 1$. By Lemma 1.8.1 we know that there is a prime number p dividing N (this may or may not be N itself). But p cannot be on our list above (a prime number on our list does not divide N – it leaves a remainder of 1 by Theorem 1.2.1). This means that from any finite list of prime numbers, we can prove the existence of a prime number not on the list: so, there are infinitely many prime numbers. $\square$

One may even prove that the sum

$$\frac{1}{2} + \frac{1}{3} + \frac{1}{5} + \frac{1}{7} + \frac{1}{11} + \cdots$$

of reciprocal prime numbers is infinite (this is one of the many proofs that there are infinitely many prime numbers). A twin prime is a prime number p such that $p + 2$ (or $p - 2$) is a prime number. Here is a list of the first few twin primes:

$$3, 5, 7, 11, 13, 17, 19, 29, 31, \ldots$$

A long-standing conjecture is that there are infinitely many twin primes. In this connection the Norwegian mathematician V. Brun (1885–1978) proved that the sum

$$B = \frac{1}{3} + \frac{1}{5} + \frac{1}{5} + \frac{1}{7} + \frac{1}{11} + \frac{1}{13} + \cdots$$

of reciprocals of the twin primes is finite! The number $B(= 1.90216 \cdots)$ is called Brun's constant. It has been computed to a high degree of accuracy by the American mathematician T. Nicely. In the latter half of 1994 Nicely discovered a disagreement between a computed and a published value of $\pi(20 \cdot 10^{12})$, where $\pi(x)$ is the number of prime numbers $\leq x$. After a long-winded process eliminating all kinds of errors, this led to the discovery[1] of the infamous FDIV bug in Intel's initial launch of their Pentium processor.

A crucial property of prime numbers (even though it looks strange at the beginning) is the following lemma.

---

[1] See http://www.trnicely.net/pentbug/pentbug.html

Lemma 1.8.3   *Let p be a prime number and suppose that $p \mid ab$, where $a, b \in$ $\mathbb{Z}$. Then $p \mid a$ or $p \mid b$.*

Proof.   If $p \nmid a$ then $\gcd(p, a) = 1$ and therefore $p \mid b$ by Corollary 1.5.10. Similarly if $p \nmid b$ then $p \mid a$. This shows that $p \mid a$ or $p \mid b$.   □

Remark 1.8.4   Lemma 1.8.3 extends to products with more than two factors: if p is a prime number and $p \mid a_1 a_2 \cdots a_n$ then $p \mid a_1$ or $p \mid a_2$ or ... or $p \mid a_n$. Can you prove this?

1.8.2   Unique factorization

We know that every number can be written as a product of prime numbers. Gauss was the first to see a potential problem hidden in this statement. Can one have two different collections

$$p_1 \leq p_2 \leq \cdots \leq p_r \quad \text{and} \quad q_1 \leq q_2 \leq \cdots \leq q_s$$

of prime numbers such that $p_1 p_2 \cdots p_r = q_1 q_2 \cdots q_s$? A bit of experimentation shows that one seems to get different numbers given different collections of prime numbers (for example $2 \cdot 3 \cdot 11 \neq 5 \cdot 13$). This is a mathematical statement crying out for a rigorous proof. Many mathematicians before Gauss took "unique factorization" for granted. Commenting on this Gauss wrote ([11], Section II)

> However, we did not wish to omit it (*the proof of unique factorization*) because many modern authors have offered up feeble arguments in place of proof or have neglected the theorem completely ...

The idea behind the proof of unique factorization is quite easy. Suppose we wish to prove that $2 \cdot 3 \cdot 11 \neq 5 \cdot 13$ without multiplying. Assume that $2 \cdot 3 \cdot 11 = 5 \cdot 13$. Then $2 \mid 5 \cdot 13$. Lemma 1.8.3 implies that $2 \mid 5$ or $2 \mid 13$. This is a contradiction.

Theorem 1.8.5   *Every non-zero natural number n can be factored uniquely into a product of prime numbers (up to changing the order of the factors):*

$$n = p_1 \cdots p_r.$$

Proof.   We may assume that $n > 1$ (since 1 factors uniquely into the empty product of prime numbers). Suppose that

$$n = p_1 \cdots p_r = q_1 \cdots q_s$$

are prime factorizations. If a prime factor p_j appears on the right hand side among $q_1, \ldots, q_s$, then we divide both sides by p_j. We can therefore assume from the beginning that the left and right hand sides of the above equation have no prime factors in common. Furthermore, we may assume that $r \geq 1$ and $s > 1$. But we know that $p_1 \mid n$ and, so by Lemma 1.8.3 (applied $s - 1$ times), we get $p_1 \mid q_1$ or $p_1 \mid q_2$ or $\ldots$ or $p_1 \mid q_s$. Assume that $p_1 \mid q_j$. The only way this can happen is if $p_1 = q_j$, and this contradicts the fact that every common prime factor has been cancelled. $\qquad\square$

There is a very nice and short proof of unique factorization using that $\mathbb{N}$ is well ordered (see Exercise 1.31). The above proof, however, seems to be the "natural" one as it carries over to more general settings.

Remark 1.8.6 Suppose that $n > 1$ is a natural number with the prime factorization

$$n = p_1^{e_1} \cdots p_r^{e_r},$$

where $e_1, \ldots, e_r \geq 0$. Then Theorem 1.8.5 shows that

$$\mathrm{div}(n) = \left\{ p_1^{k_1} \cdots p_r^{k_r} \mid 0 \leq k_1 \leq e_1, \ldots, 0 \leq k_r \leq e_r \right\}.$$

Suppose that

$$m = p_1^{f_1} \cdots p_r^{f_r},$$

where $f_1, \ldots, f_r \geq 0$. Then

$$\mathrm{div}(m) = \left\{ p_1^{k_1} \cdots p_r^{k_r} \mid 0 \leq k_1 \leq f_1, \ldots, 0 \leq k_r \leq f_r \right\}$$

and $\mathrm{div}(m) \cap \mathrm{div}(n)$ is

$$\left\{ p_1^{l_1} \cdots p_r^{l_r} \mid 0 \leq l_1 \leq e_1, \, 0 \leq l_1 \leq f_1, \, \ldots, 0 \leq l_r \leq e_r, \, 0 \leq l_r \leq f_r \right\}$$
$$= \left\{ p_1^{l_1} \cdots p_r^{l_r} \mid 0 \leq l_1 \leq \min(e_1, f_1), \, \ldots, 0 \leq l_r \leq \min(e_r, f_r) \right\}.$$

Therefore

$$\gcd(m, n) = p_1^{\min(e_1, f_1)} \cdots p_r^{\min(e_r, f_r)}.$$

Similarly, the smallest natural number having both m and n as divisors must be

$$p_1^{\max(e_1, f_1)} \cdots p_r^{\max(e_r, f_r)}.$$

This number is denoted $\mathrm{lcm}(m, n)$ and is called the *least common multiple* of m and n. So if you have access to the prime factorizations of m and n it is easy to read off the greatest common divisor and the least common multiple. Take $n = 140$ and $m = 154$. Here

$$n = 2^2 \cdot 5^1 \cdot 7^1 \cdot 11^0,$$
$$m = 2^1 \cdot 5^0 \cdot 7^1 \cdot 11^1$$

Therefore $\gcd(140, 154) = 2^1 \cdot 7^1 = 14$ and $\text{lcm}(140, 154) = 2^2 \cdot 5^1 \cdot 7^1 \cdot 11^1 = 1540$.

1.8.3 How to compute $\varphi(n)$

So far the most effective way known of computing $\varphi(n)$ for a natural number n is by way of its prime factorization. By Proposition 1.7.1

$$\varphi(n) = \varphi\left(p_1^{r_1}\right) \cdots \varphi\left(p_s^{r_s}\right)$$

where $n = p_1^{r_1} \cdots p_s^{r_s}$ is the prime factorization ($p_i \neq p_j$ for $i \neq j$) of n. So we need to know how to compute $\varphi(p^m)$ for a power p^m of a prime number p. Fortunately this is easy. First observe that a number x is relatively prime to p^m if and only if $p \nmid x$ (why?). So the natural numbers less than p^m that are not relatively prime to p^m are simply the multiples of p. We list them below:

$$0, \ p, \ 2p, \ \ldots, \ (p-1)p, \ p^2, \ \ldots, \ (p^2 - 1)p, \ p^3, \ \ldots, \ (p^{m-1} - 1)p.$$

There are p^{m-1} natural multiples of p less than p^m. This implies that $\varphi(p^m) = p^m - p^{m-1}$. Therefore

$$\varphi(n) = \left(p_1^{r_1} - p_1^{r_1-1}\right) \cdots \left(p_s^{r_s} - p_s^{r_s-1}\right) = n \left(1 - \frac{1}{p_1}\right) \cdots \left(1 - \frac{1}{p_s}\right).$$

The fact that this is the only efficient way known of computing φ gives the security underlying the RSA cryptosystem.

1.9 RSA explained

Let us return to the setting of the introduction to this chapter, where the RSA cryptosystem was described. Recall that a person wishing to receive an encrypted message must make two natural numbers e and N public. The number N (the public key) is the product of two distinct prime numbers p and q. A person wishing to send the number X ($0 \leq X < N$) sends the encrypted number $[X^e]$. Now the receiver can read this message because he knows a secret number d such that $[[X^e]^d] = X$. Here remainders are with respect to N.

We will now see how to construct the numbers e and d. By earlier results (Proposition 1.3.2 and (1.2)) we know that $[[X^e]^d] = [X^{ed}] = X$ if and only if $X \equiv X^{ed} \pmod{N}$. We also know that $\varphi(N) = \varphi(p)\varphi(q) = (p-1)(q-1)$ by subsection 1.8.3. The following proposition captures the algebraic essence of the RSA cryptosystem.

Proposition 1.9.1 *Let X be any integer and k a natural number. Then*

$$X^{k(p-1)(q-1)+1} \equiv X \pmod{N}.$$

Proof. By Corollary 1.5.11(i) it is enough to prove that

$$X^{k(p-1)(q-1)+1} \equiv X \pmod{p},$$
$$X^{k(p-1)(q-1)+1} \equiv X \pmod{q}.$$

We will prove the congruence for p (the proof for q is similar). If $p \mid X$ then $X \equiv 0 \pmod{p}$. Therefore $X^{k(p-1)(q-1)+1} \equiv 0 \pmod{p}$ and $X^{k(p-1)(q-1)+1} \equiv X \pmod{p}$. However, if $p \nmid X$, then $\gcd(X, p) = 1$. Therefore Theorem 1.7.2 gives $X^{\varphi(p)} = X^{p-1} \equiv 1 \pmod{p}$, and then we compute with congruences:

$$X^{k(p-1)(q-1)} \equiv (X^{p-1})^{k(q-1)} \equiv 1 \pmod{p}.$$

Multiplying the congruence with X, we get the desired result $X^{k(p-1)(q-1)+1} \equiv X \pmod{p}$. $\qquad\square$

1.9.1 Encryption and decryption exponents

Now we come to the selection of the encryption exponent. This exponent e is chosen as a natural number relatively prime to $\varphi(N) = (p-1)(q-1)$. Once e is chosen the decryption exponent d may be computed as follows. According to Lemma 1.5.7 we can find integers λ and μ, such that

$$\lambda(p-1)(q-1) + \mu e = 1,$$

where we may assume that $0 < \mu < (p-1)(q-1)$ (see Exercise 1.13) and therefore that $\lambda < 0$. The decryption exponent is $d = \mu$. This gives the existence of natural numbers k and d ($k = -\lambda$ and $d = \mu$) such that $k(p-1)(q-1) + 1 = de$. By (1.2) we get

$$[[X^e]^d] = [X^{ed}] = [X^{k(p-1)(q-1)+1}] = X$$

for every natural number $0 \le X < N$, where the last equality comes from Propositions 1.9.1 and 1.3.2. This is exactly the statement that the decryption of the encrypted text recaptures the text.

Notice the secret buried in $\varphi(N) = (p-1)(q-1)$. If we can compute $\varphi(N)$, we may compute a decryption key, given a public encryption key, using the Euclidean algorithm as above. Finding $\varphi(N)$ in this case is just as hard as factoring N (see Exercise 1.38). Knowledge of N and the exponents e and d is enough to "guess" the prime numbers p and q. It is therefore not safe to let different people share the same public key N (see Exercise 1.40 (HOF)).

One very practical question remains. The public key N must be the product of two enormous prime numbers p and q (more than 100 digits each). How do we find huge prime numbers with more than 100 digits without factoring numbers? The answer lies in an old result of Fermat dating back to 1640.

1.9.2 Finding astronomical prime numbers

A corollary of Euler's theorem (Theorem 1.7.2) says that a prime number p divides $a^p - a$ for all integers a. This result is due to Fermat (1601–65). In a letter dated 18 October 1640 to Frénicle de Bessy, Fermat writes

> It seems to me after this that I should tell you the foundation on which I support the demonstrations of all which concerns geometric progressions, namely: Every prime number measures infallibly one of the powers minus unity in any progression, and the exponent of this power is a divisor of the given prime number minus one; and after one has found the first power which satisfies the condition, all those whose exponents are multiples of the first satisfy the condition.

Fermat writes in his letter " ... I would send you the demonstration, if I did not fear its being too long." The first known proof of Fermat's result dates back to Euler in 1736. Later, in 1760, Euler gave his general result, which we proved in Theorem 1.7.2.

Corollary 1.9.2 (Fermat's little theorem)   *Let p be a prime number and a an integer with $\gcd(a, p) = 1$. Then*

$$a^{p-1} \equiv 1 (\mathrm{mod}\ p).$$

Proof.   This is a consequence of Theorem 1.7.2 since $\varphi(p) = p - 1$.   □

You may wonder what Fermat's big theorem is. This is known by the name "Fermat's last theorem" and goes back to 1647. It says that the equation

$$X^n + Y^n = Z^n$$

has no solutions $X, Y, Z \in \mathbb{Z}$ for $n > 2$ apart from the trivial ones where one of X, Y, Z is zero (when $n = 2$ there are infinitely many non trivial solutions). Fermat conjectured this in his notes in the margin of his copy of Diophantus's *Arithmetica* with the famous remark "For this I have discovered a truly wonderful proof, but the margin is too small to contain it." As you may know Fermat's last theorem haunted mathematicians for more than 300 years before it was finally proved by A. Wiles in 1994.

It is unlikely that Fermat could have forseen that his little theorem would play a crucial role in generating large prime numbers for use in the modern information age. By using congruences it is easy to see (since $5 \equiv -1 \pmod 6$) that

$$5^5 \equiv 5 \not\equiv 1 \pmod 6.$$

Thus by Corollary 1.9.2, 6 is not a prime number. This is of course a complicated way of proving the latter, but in fact it contains the idea for some beautiful algorithms for deciding whether a number is composite without ever trying to factor it. However,

$$8^8 \equiv (-1)^8 = 1 \pmod 9.$$

Here Corollary 1.9.2 does not tell us that 9 is composite. We are led to the following definition.

Definition 1.9.3 Let N be a composite natural number and a an integer. Then N is called a *pseudoprime* relative to the base a if $a^{N-1} \equiv 1 \pmod N$.

Notice that if the base a is not relatively prime to N then N cannot be a pseudoprime relative to a (see Exercise 1.41). A natural question is whether there exist numbers pseudoprime to every relatively prime base. The answer is yes, and the smallest example is $N = 561 = 3 \cdot 11 \cdot 17$ (see Exercise 1.45). Numbers having this property are called Carmichael numbers (or pseudoprimes). It was proved recently [1] that there are infinitely many Carmichael numbers.

We are left with the fact that there are composite numbers that are not distinguished from prime numbers by Corollary 1.9.2. There is a surprisingly simple way to improve this situation. The key point is the following lemma.

Lemma 1.9.4 *Let p be a prime number and $x \in \mathbb{Z}$. If $x^2 \equiv 1 \pmod p$ then $x \equiv \pm 1 \pmod p$.*

Proof. By assumption, $p \mid x^2 - 1 = (x + 1)(x - 1)$. Thus, by Lemma 1.8.3, $p \mid x + 1$ or $p \mid x - 1$. This completes the proof. $\square$

Consider, say, $N = 341$. Using repeated squaring we compute

$$2^{340} \equiv 1 \,(\text{mod } 341).$$

From this we cannot deduce that 341 is composite. But using Lemma 1.9.4 we can drag 2 through some more questioning that tell us whether 341 really is composite. Assuming, then, that 341 is a prime number, Lemma 1.9.4 gives that

$$2^{170} \equiv \pm 1 \,(\text{mod } 341),$$

since $(2^{170})^2 = 2^{340}$. Again one computes that $2^{170} \equiv 1 \,(\text{mod } 341)$. Now we reach the crucial question. Since $2^{170} = (2^{85})^2$, Lemma 1.9.4 implies that $2^{85} \equiv \pm 1 \,(\text{mod } 341)$. In this step, 2 breaks down and tells us that

$$2^{85} \equiv 32 \,(\text{mod } 341)$$

and therefore that 341 cannot be a prime number. From this example we get the following definition.

Definition 1.9.5   An odd composite number N is called a *strong pseudoprime* relative to the base a if either $a^q \equiv 1 \,(\text{mod } N)$ or there exists $i = 0, \ldots, k - 1$ such that

$$a^{2^i q} \equiv -1 \,(\text{mod } N),$$

where $N - 1 = 2^k q$ and $2 \nmid q$.

The strong pseudoprimes are precisely the composite numbers, which pass both tests (Corollary 1.9.2 and Lemma 1.9.4) without getting caught. The following result shows that a number that fails repeated application of Lemma 1.9.4 (as for $N = 341$ and $a = 2$) must be a composite number.

Proposition 1.9.6   *Let p be an odd prime number and suppose that*

$$p - 1 = 2^k q,$$

where $2 \nmid q$. If $a \in \mathbb{Z}$ and $\gcd(a, p) = 1$ then either $a^q \equiv 1 \,(\text{mod } p)$ or there exists $i = 0, \ldots, k - 1$ such that

$$a^{2^i q} \equiv -1 \,(\text{mod } p).$$

Proof.   Let $a_i = a^{2^i q}$, $i = 0, \ldots, k$. Observe that $a_k \equiv 1 \,(\text{mod } p)$ by Corollary 1.9.2 and that $a_{i+1} = a_i^2$ for $i = 0, \ldots, k - 1$. Therefore $a_0 \equiv 1 \,(\text{mod } p)$ if and only if $a_i \equiv 1 \,(\text{mod } p)$ for every $i = 0, \ldots, k$. So, if $a_0 \not\equiv 1 \,(\text{mod } p)$ then there exists a_i, $i \geq 0$, such that $a_i \not\equiv 1 \pmod{p}$. Let j be the largest index with this property. Since $j < k$ and $a_j^2 \equiv a_{j+1} \equiv 1 \pmod{p}$ we get $a_j \equiv -1$ $\pmod{p}$ by Lemma 1.9.4. $\qquad\square$

The reason strong pseudoprimes are extremely useful in real-life primality testing is the following theorem, due to M. Rabin [20].

Theorem 1.9.7 (Rabin)   *Suppose that $N > 4$ is an odd composite integer and let B be the number of bases a $(1 < a < N)$ such that N is a strong pseudoprime relative to a. Then*

$$ B < \varphi(N)/4 \leq (N-1)/4. $$

Theorem 1.9.7 shows the strong contrast between a strong pseudoprime and a pseudoprime to a base a. There are true pseudoprimes (composite numbers pseudoprime to every relatively prime base). Theorem 1.9.7 states that we can find many bases revealing that a given composite number is not a prime number! Suppose that we are given a natural number N and a randomly chosen a, $1 < a < N$. If N is composite then the probability that N is a strong pseudoprime relative to a is $< 1/4$ by Theorem 1.9.7. If we have a good method of generating (uniformly distributed) random numbers,[2] then we can try out a sequence of random bases $1 < a_1, \ldots, a_m < N$. The upshot is that if N is a strong pseudoprime relative to the m random bases $a_1, \ldots, a_m$ then the probability that N is composite is less than

$$ (1/4)^m. $$

In fact the probability is usually much smaller. For example, if a number p of around 180 digits (600 bits) is tested and p is a strong pseudoprime to just one base then the probability that p is composite is less than $(1/2)^{76}$. Already for $m \geq 30$ the rough estimate $(1/4)^m$ is comparable to the probability of a hardware error in your computer caused by cosmic radiation (quoting Knuth). So if a number is a strong pseudoprime relative to more than 30 random bases then the number tested is a prime number for all practical purposes. This is basically how one builds huge prime numbers for use in cryptography. Starting with a random integer x (with more than 100 digits), one searches for the first probable prime number $\geq x$.

---

[2] The best sources of randomness are atmospheric noise from a radio (http://www.random.org) or radioactive decay (http://www.fourmilab.ch/hotbits/).

1.10 Algorithms for prime factorization

One way of breaking the RSA encryption is by having effective algorithms for prime factorization. So far these work only up to 155 digits, and the largest number took six months to factor using distributed supercomputing over the internet. In August 1999 RSA-155 was factored. Here is part of the press release from RSA Labs:

> Factoring the 512-binary-bit key, equivalent to 155 decimal digits and called RSA-155, took the team a total elapsed time of 5.2 months, not including nine weeks needed for preliminary computations, and was accomplished using 292 individual computers located at 11 different sites in The Netherlands, Canada, the United Kingdom, France, Australia and the United States. Prior to this, the largest RSA key length to be factored was 140 decimal digits long in February of this year. RSA's recommended key lengths are 230 digits or more. ... These latest results were achieved using about 160 175-400-MHz SGI and Sun workstations, eight 250-MHz SGI Origin 2000 processors, 120 300-450-MHz Pentium II PCs and four 500-MHz Digital/Compaq CPUs, and required approximately 8000 MIPS-years of CPU effort. The specific approach used to determine the prime factors was based on the work done to solve the RSA-140 Challenge earlier this year.

The statement that every integer can be written as a product of prime numbers is a typical mathematical statement with a simple proof. Things become much more complicated when you (inspired by Gauss) ask for a good algorithm for factoring a given integer N. In a non-trivial factorization $N = ab$ one of the factors a and b must be $\leq \sqrt{N}$. If N is even, 2 divides it and we have found a factor. If N is odd we may find a factor of N by starting with 3 and trying division by odd numbers up to $\sqrt{N}$. This procedure is called *trial division*. The number of steps in trial division is proportional to the size of the smallest prime factor. This is extremely slow. If you want to factor a 100-digit number that is the product of two 50-digit prime numbers, you must carry out approximately 10^{50} steps of trial division. If every step takes 10^{-10} seconds, you will have to wait for 10^{40} seconds (or approximately 10^{32} years). It is not clear, though, that there are better algorithms. In fact the three faster algorithms we will describe each contain an ingenious idea. They are all tied to the Euclidean algorithm. The object is to hunt down a number $a \in \mathbb{N}$ such that $1 < \gcd(a, N) < N$, where N is the composite number we wish to factor.

1.10.1 The birthday problem

Suppose that N people are gathered in an auditorium. What is the probability that two of them share the same birthday? This is a problem easily solved by elementary probability theory. Consider the "inverse" problem: what is the probability, $P(N)$, that none of them share a birthday? We get, for example, that

$$P(2) = \frac{364}{365}$$

since there are 364 possible dates left when one is taken. Similarly,

$$P(3) = \frac{364}{365} \cdot \frac{363}{365}.$$

In general,

$$P(N) = \frac{365 \cdot 364 \cdots (365 - N + 1)}{365^N}.$$

At $N = 23$, $P(N)$ is already less than 0.5. So if there are more than 23 people present there is more than a 50% chance that two share the same birthday. If there are 50 people present there is more than a 97% chance that two share the same birthday.

The mathematical abstraction is sampling with replacement from a sample space consisting of N objects. The average number of samplings before a repetition occurs can be computed as the mean value of a stochastic variable. When N is big this mean value is close to

$$\sqrt{\frac{\pi N}{2}}.$$

1.10.2 Pollard's ρ-algorithm

How does the birthday problem relate to the factoring of a composite integer N? Suppose that p is a prime number dividing N and that we are given two numbers $0 \leq a, b < N$ with $a \equiv b \pmod{p}$. Then $p \mid a - b$. Therefore $1 < \gcd(a - b, N) \leq N$ and, if $a \neq b$, $\gcd(a - b, N)$ is a non-trivial factor in N. This innocent observation contains the idea for a much faster factoring algorithm than trial division. Suppose we have a way of generating random integers $X_1, X_2, \ldots$ with $0 \leq X_i < N$. We know by subsection 1.10.1 that on average we see $\sqrt{(\pi N)/2}$ random numbers before a repetition occurs. For factoring purposes it suffices to have a repetition modulo p, where p is the smallest prime dividing N: if $X_i \equiv X_j \pmod{p}$ then $1 < \gcd(X_i - X_j, N) \leq N$. So it is sufficient to look at the random integers $X_1, X_2, \ldots$ modulo p. Here we only see $\sqrt{(\pi p)/2}$ random numbers, on average, before a repetition occurs.

It is not easy to generate true random numbers in mathematics. Let us rely on a function that generates a sequence of numbers conjectured to be sufficiently random. Consider the function from $\mathbb{Z}/N$ to $\mathbb{Z}/N$ given by

$$f(X) = [X^2 + 1]_N. \qquad (1.5)$$

Start out with $X_0 = 0$ and let $X_{i+1} = f(X_i)$ in each successive step. This sequence will contain repetitions modulo p (there are only p remainders so there will be a repetition when $i \geq p$). How do we check for repetitions modulo p? The following lemma gives the crux of the algorithm.

Lemma 1.10.1 *Let $f : M \to M$ be a function where M is a finite set. Pick $x_0 \in M$ and generate the sequence $x_0, x_1, x_2, \ldots$, where $x_{i+1} = f(x_i)$ for $i \geq 0$. There exist $i, j \in \mathbb{N}$ such that $i \neq j$ and $x_i = x_j$. Furthermore there exists $n > 0$ such that $x_n = x_{2n}$. The sequence $y_0, y_1, y_2, \ldots$ given by $y_0 = x_0$ and $y_{i+1} = f(f(y_i))$ for $i \geq 0$ equals the sequence $x_0, x_2, x_4, \ldots$.*

Proof. We have a map $g : \mathbb{N} \to M$ given by $g(n) = f^n(x_0)$. Since M is a finite set, g cannot be injective. Thus there exist $i, j \in \mathbb{N}$ with $i \neq j$ such that $g(i) = g(j)$. This shows that $x_i = x_j$ for $i \neq j$.

Suppose that $x_i = x_j$ for $j > i$. If $n \geq i$ and $2n = n + k(j - i)$ with $k \geq 0$ we must have $x_n = x_{2n}$. So choosing $k \geq 0$ such that $n = k(j - i) \geq i$ gives the desired n. As $x_{m+2} = f(f(x_m))$ it follows that $y_m = x_{2m}$.   □

Now we have all the tools for building a factoring algorithm based on recognizing repetitions modulo p. The key point is the existence of a repetition modulo p of the form $X_n \equiv X_{2n} \pmod{p}$, as pointed out in Lemma 1.10.1.

We start out by putting $X_0 = Y_0 = 0$. At each step we iterate $X_{i+1} = f(X_i)$ and $Y_{i+1} = f(f(Y_i))$ using the function f in (1.5). Then we compute $d = \gcd(Y_{i+1} - X_{i+1}, N)$ using the Euclidean algorithm. If d equals 1 or N we repeat the process. If not, d must be a non-trivial factor in N and we are done. An example is given below.

Example 1.10.2 Let $N = 11 \cdot 13 = 143$. Then

i	0	1	2	3	4	5	6	7	8	9
X_i	0	1	2	5	26	105	15	83	26	105
Y_i	0	2	26	15	26	15	26	15	26	15

The X_i-sequence turns into the sequence $0, 1, 2, 5, 4, 6, 4, 6, 4, \ldots$ viewed modulo 11. At the sixth step above $Y_6 - X_6 = 11$ and the factor 11 is found.

Of course there is a problem with this algorithm if repetition modulo N coincides with repetition modulo p. This is rather unlikely for large N since the sequence modulo N repeats after $\sqrt{(\pi N)/2}$ steps on average compared with $\sqrt{(\pi p)/2}$ steps modulo p on average.

This algorithm for factoring is called Pollard's ρ-algorithm (because ρ represents the shape of the sequence repeating itself). It was invented in 1975 by J. M. Pollard. The Pollard ρ-algorithm needs $\sqrt[4]{N}$ steps on average for factoring an integer N as compared with $\sqrt{N}$ steps for trial division. We move on to describe another factoring algorithm due to Pollard.

1.10.3 Pollard's $(p-1)$-algorithm

Suppose we wish to factor a composite number N divisible by a prime number p. If a is an integer and $p \nmid a$ then

$$a^{p-1} \equiv 1 \,(\text{mod } p)$$

by Corollary 1.9.2. If m is a natural number such that $p - 1 \mid m$ then

$$a^m \equiv 1 \,(\text{mod } p). \tag{1.6}$$

So if we have a and m such that (1.6) holds, we may conclude that $\gcd(N, a^m - 1) > 1$ since $p \mid a^m - 1$. This suggests that a good strategy for finding a non-trivial factor of N proceeds by systematically trying out a and m in the hope that they fit (1.6). We can use the Euclidean algorithm to compute

$$d = \gcd(N, a^m - 1) = \gcd(N, [a^m - 1]_N).$$

Computing $[a^m - 1]_N$ can be done using repeated squaring, first evaluating $[a^m]_N$. If $1 < d < N$ we have found a non-trivial factor of N. If $d = 1$ then we try with a different (bigger) m. If $d = N$ then m might work with a different a. This is the idea behind Pollard's $(p - 1)$-algorithm. This algorithm is very successful if N contains a prime factor p such that $p - 1$ is a product of small primes. In fact one builds up m as a product of primes of increasing size. The jackpot in the algorithm occurs when we hit an m that is divisible by $p - 1$, where p is a prime factor of N.

How do we search systematically through the m-values for a specific a? A good strategy is to decide on a bound B, considering only the prime numbers $< B$ dividing m. One then takes the powers of the prime number $q < B$ entering m as the least integer greater than or equal to $\log_q \sqrt{N}$ ([5], Section 4.3). Again we will do a toy example that fully explains the basic idea.

Example 1.10.3   Consider $N = 143 = 11 \cdot 13$ and $a = 2$. We consider primes $< B = 5$. Thus in this case $m = 2^4 3^3 = 432$. Using repeated squaring we find that

$$[2^{432}]_{143} = 92.$$

Therefore

$$\gcd(143,\, 2^{432} - 1) = \gcd(143, 91) = 13.$$

We have found the factor 13 in 143.

To protect a public RSA-key from the Pollard $(p-1)$-algorithm one should choose (secret) prime numbers p and q such that $p-1$ and $q-1$ are not products of small prime numbers.

The (hidden group theoretic) idea behind the $(p-1)$-algorithm can be used to construct a much stronger factoring algorithm using arithmetic on elliptic curves (this was done by Lenstra in 1985).

1.10.4   The Fermat–Kraitchik algorithm

Currently the most effective algorithms for factoring difficult RSA integers originate in the historic fact that if an integer N can be written as the difference $x^2 - y^2$ between two squares, we have the factorization $N = x^2 - y^2 = (x + y)(x - y)$. However, if an odd number $N = uv$ is composite then

$$N = \left(\frac{u+v}{2}\right)^2 - \left(\frac{u-v}{2}\right)^2.$$

This method of factoring goes back to Fermat. Suppose we wish to factor N. Fermat's method uses the function

$$S(x) = x^2 - N$$

and a search for x such that $S(x)$ is a square. Usually one runs through $x = [\sqrt{N}], [\sqrt{N}] + 1, \ldots$, where $[\sqrt{N}]$ denotes the largest integer $\leq \sqrt{N}$. Putting $N = 2491$, one finds $S(49) = -90$, $S(50) = 9 = 3^2$. This means that $2491 = (50 + 3)(50 - 3) = 53 \cdot 47$. Of course, using this method on a composite number such as 2^{1000} works just as poorly as trial division. There is a beautiful variation of Fermat's method, due to M. Kraitchik (1882–1957), using congruences. The insight is that to find a factor of N it usually suffices that N divides $x^2 - y^2$. If

$$N \mid x^2 - y^2 = (x + y)(x - y)$$

and N does not divide either $x + y$ or $x - y$ then we may conclude that $\gcd(x + y, N) > 1$ by Corollary 1.5.10 and use the Euclidean algorithm to find $\gcd(N, x + y)$, which is a non-trivial factor of N. So one should look for integers x, y such that

$$x^2 \equiv y^2 \,(\mathrm{mod}\ N),$$
$$x \not\equiv \pm y \,(\mathrm{mod}\ N).$$

Suppose that we have collected $x_1, \ldots, x_n$ along with the congruences

$$x_1^2 \equiv a_1 \,(\mathrm{mod}\ N), \quad \ldots, \quad x_n^2 \equiv a_n \,(\mathrm{mod}\ N)$$

for some integers $a_1, \ldots, a_n$. If a subset $a_{i_1}, \ldots, a_{i_r}$ of $a_1, a_2, \ldots, a_n$ satisfies that $a_{i_1} \cdots a_{i_r}$ is a square then

$$(x_{i_1} \cdots x_{i_r})^2 \equiv a_{i_1} \cdots a_{i_r} \,(\mathrm{mod}\ N)$$

by Proposition 1.3.4 and we have our congruence $x^2 \equiv y^2 \,(\mathrm{mod}\ N)$. This congruence may or may not satisfy $x \not\equiv \pm y \,(\mathrm{mod}\ N)$. To tell whether a number n is square we factor it,

$$n = p_1^{m_1} \cdots p_r^{m_r},$$

using some predefined factor basis $P = \{p_1, \ldots, p_r\}$ of (small) prime numbers. In this context, n is a square if and only if all the exponents $m_1, \ldots, m_r$ are even. Using linear algebra there is a method of systematically finding a subset $\{i_1, \ldots, i_r\}$ such that $a_{i_1} \cdots a_{i_r}$ is a square (see Exercise 1.49).

Let us apply this algorithm to the numbers we get from the function $S(x) = x^2 - N$. Notice that $x^2 \equiv S(x) \,(\mathrm{mod}\ N)$. For $N = 2041$ (this example is from [19]) we illustrate this in the table below. The marked entries together indicate the subset whose product is a square.

x	$S(x)$	Factorization	Marked
46	75	$3 \cdot 5^2$	✓
47	168	$2^3 \cdot 3 \cdot 7$	✓
48	263	263	
49	360	$2^3 \cdot 3^2 \cdot 5$	✓
50	459	$3^3 \cdot 17$	
51	560	$2^4 \cdot 5 \cdot 7$	✓

The above table shows that $S(46)S(47)S(49)S(51) = 75 \cdot 168 \cdot 360 \cdot 560 = (2^5 \cdot 3^2 \cdot 5^2 \cdot 7)^2$ is a square. Putting $u = 2^5 \cdot 3^2 \cdot 5^2 \cdot 7$, we get

$$u^2 = 50400^2 \equiv 1416^2 \,(\mathrm{mod}\ 2041).$$

Now we know that $u^2 \equiv v^2 \pmod{2041}$ where $v = 46 \cdot 47 \cdot 49 \cdot 51 = 5402838 \equiv 311 \pmod{2041}$. Using the Euclidean algorithm one finds the greatest common divisor of $u - v = 1416 - 311 = 1105$ and 2041, which is 13. We have found the factorization $2041 = 13 \cdot 157$. Using the original method of Fermat we would have to wait until $x = 85$ before $S(x)$ is a square. The heavy part of the algorithm is factoring $S(x) = x^2 - N$. Around 1982, Pomerance discovered a nice trick that avoids this. His observation was that a prime power p^r divides $S(x)$ if and only if it divides $S(x + kp^r)$, where $k \in \mathbb{Z}$. So if we can locate a number x such that $p^r \mid S(x)$ then we know in advance that $p^r \mid S(x + p^r)$, $S(x + 2p^r)$, ... This is a so-called sieving procedure (like the sieve of Eratosthenes, which eliminates multiples of prime numbers). It leads to a factorization algorithm called the quadratic sieve. In [19] you can find a nice description of this and more advanced sieving methods for factoring. These are currently the most effective algorithms for the challenges issued by RSA Labs. In fact RSA-155 was factored using sieving.

1.11 Quadratic residues

In this section we introduce the fundamentals of quadratic residues modulo a prime number p. Gauss originally developed this theory, starting with the question whether a number a has a square root modulo p: can one find an integer x such that $x^2 \equiv a \pmod{p}$? This question led Gauss to exceptionally beautiful mathematics (see "Congruences of the second degree," Section IV in [11] and be sure to enjoy the clarity of the exposition).

Later, we will use quadratic residues when writing prime numbers $\equiv 1 \pmod 4$ as a sum of two squares.

Definition 1.11.1 Let p be a prime number. If $p \nmid a$ then a is called a *quadratic residue modulo* p if it is congruent to a square modulo p (i.e. there exists $x \in \mathbb{Z}$ such that $a \equiv x^2 \pmod{p}$). Otherwise a is called a *quadratic non-residue modulo* p. If $p \mid a$ then a is considered neither a quadratic residue nor a quadratic non-residue. This definition is contained in the Legendre symbol

$$\left(\frac{a}{p} \right) = \begin{cases} 0 & \text{if } p \mid a, \\ 1 & \text{if } a \text{ is a quadratic residue modulo } p, \\ -1 & \text{if } a \text{ is a quadratic non-residue modulo } p. \end{cases}$$

If $a \equiv x^2 \pmod{p}$ for some integer $x \in \mathbb{Z}$, we may find a y such that $0 \leq y < p$ and $a \equiv y^2 \pmod{p}$. We simply put $y = [x]_p$. Then $y \equiv x \pmod{p}$ and therefore $y^2 \equiv x^2 \pmod{p}$. Thus the quadratic residues among the numbers $1, 2, \ldots, p - 1$ are the numbers

$$[1^2], [2^2], [3^2], \ldots, [(p - 1)^2],$$

where the remainder is with respect to p. This is reflected in the Legendre symbol:

$$\left(\frac{a}{p}\right) = \left(\frac{a + kp}{p}\right)$$

where $k \in \mathbb{Z}$.

Example 1.11.2 Let $p = 7$. Since a non-zero square x^2 modulo p is always congruent to one of the squares $1^2, 2^2, 3^2, 4^2, 5^2, 6^2$, we may list the quadratic residues among $1, 2, 3, 4, 5, 6$ as

$$[1^2] = 1,$$
$$[2^2] = 4,$$
$$[3^2] = 2,$$
$$[4^2] = 2,$$
$$[5^2] = 4,$$
$$[6^2] = 1$$

by taking the remainder after division by 7. Notice the symmetry above: $[3^2] = [4^2]$, $[2^2] = [5^2]$, $[1^2] = [6^2]$. This is a consequence of the fact that $x^2 \equiv (7 - x)^2 \pmod{7}$. There are an equal number of quadratic residues, $\{1, 2, 4\}$, and quadratic non-residues, $\{3, 5, 6\}$.

Proposition 1.11.3 *Let p denote an odd prime. Half the numbers $1, 2, 3, \ldots, p - 1$ are quadratic residues; the other half are quadratic non-residues modulo p.* [3]

Proof. We already know that the quadratic residues are $[1^2], [2^2], \ldots,$ $[(p - 1)^2]$. But since $x^2 \equiv (p - x)^2 \pmod{p}$, we see that the quadratic residues are given by the first $(p - 1)/2$ numbers $[1^2], [2^2], \ldots, [((p - 1)/2)^2]$. These numbers really are different. If $[i^2] = [j^2]$ then $i^2 \equiv j^2 \pmod{p}$ and $p \mid i^2 - j^2 = (i + j)(i - j)$. Therefore $p \mid i + j$ or $p \mid i - j$. This is only possible if $i = j$, because $0 \leq i, j \leq (p - 1)/2$. So there are $(p - 1)/2$ quadratic residues

---

[3] An interesting problem is how the quadratic non-residues are distributed among $1, 2, \ldots, p - 1$. In particular, how big is the smallest quadratic non-residue a? If a generalization of the famous Riemann hypothesis is true one can prove that $a < 2(\log p)^2$, where log denotes the natural logarithm.

and therefore $(p - 1) - (p - 1)/2 = (p - 1)/2$ quadratic non-residues among the numbers $1, 2, \ldots, p - 1$. □

The following important theorem is due to Euler.

Theorem 1.11.4 (Euler) *Let p be an odd prime and let a be an integer not divisible by p. Then*

$$\left(\frac{a}{p}\right) \equiv a^{(p-1)/2} \,(\text{mod } p).$$

Proof. If a is a quadratic residue then $a \equiv x^2 \,(\text{mod } p)$, where $p \nmid x$ for some $x \in \mathbb{Z}$. Therefore

$$a^{(p-1)/2} \equiv (x^2)^{(p-1)/2} = x^{p-1} \equiv 1 \,(\text{mod } p)$$

by Corollary 1.9.2. Therefore we have at least $(p - 1)/2$ incongruent solutions to the congruence

$$X^{(p-1)/2} - 1 \equiv 0 \,(\text{mod } p). \tag{1.7}$$

What is shown in Exercise 1.50 implies that (1.7) can have at most $(p - 1)/2$ incongruent solutions. Therefore a quadratic non-residue a cannot be a solution to (1.7). Thus $a^{(p-1)/2} \not\equiv 1 \,(\text{mod } p)$ and therefore $a^{(p-1)/2} \equiv -1 \,(\text{mod } p)$ by Lemma 1.9.4. This finishes the proof of Lemma 1.9.4. □

Corollary 1.11.5 *Let p be an odd prime. Then the Legendre symbol satisfies*

$$\left(\frac{ab}{p}\right) = \left(\frac{a}{p}\right)\left(\frac{b}{p}\right).$$

Another very nice consequence of Theorem 1.7.2 is the following.

Proposition 1.11.6 *Let p be an odd prime. Then -1 is a quadratic residue modulo p if $p \equiv 1 \,(\text{mod } 4)$ and a quadratic non-residue if $p \equiv 3 \,(\text{mod } 4)$.*

Proof. From Theorem 1.7.2 we get

$$\left(\frac{-1}{p}\right) = (-1)^{(p-1)/2}.$$

Now the result follows, since $(p - 1)/2$ is even if $p \equiv 1 \,(\text{mod } 4)$ and odd if $p \equiv 3 \,(\text{mod } 4)$. □

We move on to a celebrated lemma due to Gauss. We give the proof because it is very similar to the proof of Theorem 1.7.2. First we need some notation. Let p be an odd prime number. Then every integer $a \in \mathbb{Z}$ such that $p \nmid a$ is congruent to precisely one number (its remainder) from

$$M = \{1, 2, 3, \ldots, p - 1\}.$$

The clever thing is to break M into two and flip the right hand part below zero. We do this by replacing x by $x - p$ if $x > (p - 1)/2$. This means that every number from M is congruent to precisely one number in the set

$$S = \left\{ -\frac{p - 1}{2}, -\frac{p - 3}{2}, \ldots, -2, -1, 1, 2, \ldots, \frac{p - 3}{2}, \frac{p - 1}{2} \right\}.$$

Consider the list

$$a, 2a, 3a, \ldots, \frac{p - 1}{2}a,$$

where $p \nmid a$. None of these numbers is divisible by p (why not?). Also, they satisfy $ia \not\equiv \pm ja \pmod{p}$, since $p \nmid i - j$ and $p \nmid i + j$ when $i \neq j$ and $0 \leq i, j \leq (p - 1)/2$. This means that ia is congruent to a unique number in $\{1, 2, \ldots, (p - 1)/2\}$, up to a sign.

Definition 1.11.7   Let $\mu(a)$ denote the number of elements from the list

$$a, 2a, 3a, \ldots, \frac{p - 1}{2}a$$

congruent to a negative number in S.

Example 1.11.8   Let $p = 11$. Then

$$S = \{-5, -4, -3, -2, -1, 1, 2, 3, 4, 5\}.$$

If $a = 6$ then $6 \equiv -5, 12 \equiv 1, 18 \equiv -4, 24 \equiv 2, 30 \equiv -3$ modulo 11. This means that $\mu(a) = 3$ in this case.

Remark 1.11.9   Notice that $\mu(a)$ also is the number of elements in

$$\{[a], [a2], \ldots, [a(p - 1)/2]\} \cap \{(p + 1)/2, \ldots, p - 1\}.$$

Here we count the remainders $> (p - 1)/2$.

Lemma 1.11.10 (Gauss)   *Keep the above notation. Then*

$$\left(\frac{a}{p}\right) = (-1)^{\mu(a)}.$$

Proof. An element ja, where $j = 1, \ldots, (p-1)/2$, is congruent to $\pm m_j$, where $1 \le m_j \le (p-1)/2$. Since ia cannot be congruent to $\pm ja$ modulo p, when $i \ne j$ and $1 \le i, j \le (p-1)/2$ (this amounts to the same argument as in the proof of Theorem 1.7.2), it follows that

$$a^{(p-1)/2}\left(\frac{p-1}{2}\right)! \equiv (-1)^{\mu}\left(\frac{p-1}{2}\right)! \, (\text{mod } p).$$

Since $p \nmid ((p-1)/2)!$ we get

$$a^{(p-1)/2} \equiv (-1)^{\mu} \, (\text{mod } p)$$

and Theorem 1.11.4 finishes the proof.   □

Corollary 1.11.11   *Let p be an odd prime. Then 2 is a quadratic residue modulo p if $p \equiv 1, 7 \,(\text{mod } 8)$ and a quadratic non-residue if $p \equiv 3, 5 \,(\text{mod } 8)$.*

Proof. The number $\mu = \mu(2)$ in Lemma 1.11.10 is the number of elements in the list

$$1 \cdot 2, 2 \cdot 2, 3 \cdot 2, \ldots, \frac{p-1}{2} \cdot 2$$

that are greater than $(p-1)/2$. To compute μ we consider two cases. If $p \equiv 1 \,(\text{mod } 4)$ then

$$\mu = \frac{p-1}{2} - \frac{p-1}{4} = \frac{p-1}{4}.$$

If $p \equiv 3 \,(\text{mod } 4)$ then

$$\mu = \frac{p-1}{2} - \frac{p-3}{4} = \frac{p+1}{4}.$$

Using Lemma 1.11.10 we conclude that

$$\left(\frac{2}{p}\right) = \begin{cases} 1 & \text{if } p \equiv 1 \,(\text{mod } 8), \\ -1 & \text{if } p \equiv 3 \,(\text{mod } 8), \\ -1 & \text{if } p \equiv 5 \,(\text{mod } 8), \\ 1 & \text{if } p \equiv 7 \,(\text{mod } 8), \end{cases}$$

since $(p-1)/4$ is even when $p \equiv 1 \pmod 8$ and odd when $p \equiv 5 \pmod 8$ and $(p+1)/4$ is even when $p \equiv 7 \pmod 8$ and odd when $p \equiv 3 \pmod 8$. $\square$

Now we know how to compute $(\frac{-1}{p})$ and $(\frac{2}{p})$, but we are missing the crucial insight needed to get our hands on $(\frac{a}{p})$ in general. This insight is one of the most beautiful results in the history of mathematics. It is known as the law of quadratic reciprocity (due to Gauss, of course). It states that

$$\left(\frac{p}{q}\right)\left(\frac{q}{p}\right) = (-1)^{(p-1)(q-1)/4},$$

where p and q are odd primes. Put another way,

$$\left(\frac{p}{q}\right) = \left(\frac{q}{p}\right)(-1)^{(p-1)(q-1)/4} = \begin{cases} -\left(\dfrac{q}{p}\right) & \text{if } p \equiv q \equiv 3 \,(\text{mod } 4), \\[2ex] \left(\dfrac{q}{p}\right) & \text{otherwise.} \end{cases}$$

Think about it. If you have two odd primes p and q, it is totally unexpected that the two congruences

$$x^2 \equiv q \,(\text{mod } p) \qquad \text{and} \qquad x^2 \equiv p \,(\text{mod } q)$$

should have any connection. We will give a proof of quadratic reciprocity in Section 4.7, when we will have access to some more abstract algebra. Let us give an example showing how the Legendre symbol is computed using these rules.

Example 1.11.12

$$\left(\frac{19}{43}\right) = -\left(\frac{43}{19}\right) = -\left(\frac{5}{19}\right) = -\left(\frac{19}{5}\right) = -\left(\frac{4}{5}\right) = -\left(\frac{2}{5}\right)\left(\frac{2}{5}\right) = -1.$$

By the magic of the law of quadratic reciprocity we have proved that $x^2 \equiv 19 \,(\text{mod } 43)$ has no solutions.

1.12 Exercises

1. Prove that if a subset $S \subseteq \mathbb{Z}$ has a first element then the latter has to be unique.
2. Let $x, d \in \mathbb{Z}$, where $d > 0$. Prove that $M \cap \mathbb{N} \neq \emptyset$, where $M = \{x - qd \mid q \in \mathbb{Z}\}$.

3. Let $a, b, N \in \mathbb{Z}$, where $N > 0$. Prove that $[ab] = [[a][b]]$, where $[x]$ denotes the remainder of x after division by N.

4. Verify that the remainder of 2^{340} after division by 341 is 1, using the repeated squaring algorithm.

5. Let τ be a natural number > 1. A τ-adic expansion of a number $x \in \mathbb{N}$ is the expression

$$x = a_0 + a_1\tau + \cdots + a_r\tau^r,$$

where $r \in \mathbb{N}$, $a_i \in \mathbb{N}$ and $0 \le a_i < \tau$.

   (i) Compute a 3-adic expansion of 17.
   (ii) Prove that every $x \in \mathbb{N}\backslash\{0\}$ can be written as

$$x = a\tau^r + b,$$

   where $0 \le a < \tau, 0 \le b < \tau^r$ and $r = \max\{s \in \mathbb{N} \mid \tau^s \le x\}$.
   (iii) Prove that every natural number has a unique τ-adic expansion.

6. Let a be a number written (in base 10) as

$$a_0 \cdot 10^0 + a_1 \cdot 10^1 + a_2 \cdot 10^2 + \cdots + a_n \cdot 10^n$$

where $0 \le a_i < 10$.

   (i) Prove that 2 divides a if and only if 2 divides a_0.
   (ii) Prove that 4 divides a if and only if 4 divides $a_0 + 2a_1$.
   (iii) Prove that 8 divides a if and only if 8 divides $a_0 + 2a_1 + 4a_2$.
   (iv) Prove that 5 divides a if and only if 5 divides a_0.
   (v) Prove that 9 divides a if and only if 9 divides the sum
       $a_0 + a_1 + \cdots + a_n$ of its digits.
   (vi) Prove that 3 divides a if and only if 3 divides the sum of its digits.
   (vii) Prove that 11 divides a if and only if 11 divides

$$a_0 - a_1 + a_2 - \cdots.$$

   (viii) What is the rule for divisibility by 7?

7. Suppose that someone tricks you into believing that $233 \cdot 577 = 135441$. Use congruences to prove in a flash that this is wrong. Is there a smart way of using congruences to double-check computations such as $a + b$ and ab for integers a and b? Give a few examples.

8. Prove that $3 \mid 4^n - 1$, where $n \in \mathbb{N}$.

9. Let $m, n \in \mathbb{Z}$ not both equal zero. Prove that

$$\gcd(m, n) = \max \operatorname{div}(m) \cap \operatorname{div}(n),$$

where $\max(m, n) = m$ if $m \ge n$ and $\max(m, n) = n$ if $m < n$.

10. Let $u, v \in \mathbb{Z}$. Show that
    (i) $2 \mid u, v \Rightarrow \gcd(u, v) = 2\gcd(u/2, v/2)$.
    (ii) $2 \mid u, 2 \nmid v \Rightarrow \gcd(u, v) = \gcd(u/2, v)$.
    (iii) Use (i) and (ii) to construct a "new" Euclidean algorithm, where you also apply the fact that $\gcd(u, v) = \gcd(u - v, v)$. Give a few examples.
    The "new" Euclidean algorithm alluded to in this exercise is called the binary Euclidean algorithm. It was discovered in 1961.
11. Let $x, y, z, d \in \mathbb{Z}$. Prove the following statements.
    (i) $x \equiv x \pmod{d}$.
    (ii) If $x \equiv y \pmod{d}$ then $y \equiv x \pmod{d}$.
    (iii) If $x \equiv y \pmod{d}$ and $y \equiv z \pmod{d}$ then $x \equiv z \pmod{d}$.
12. Compute $\lambda, \mu \in \mathbb{Z}$ such that $89\lambda + 55\mu = 1$ and find all solutions $x \in \mathbb{Z}$ to

$$89x \equiv 7 \pmod{55}.$$

13. Suppose that $\lambda N + \mu M = d$, where $\lambda, \mu, M, N \in \mathbb{Z}$ and $N > 0$. Prove that one may find $\lambda', \mu' \in \mathbb{Z}$ such that

$$\lambda'N + \mu'M = d,$$

    where $0 \leq \mu' < N$.
14. Let $m, n \in \mathbb{Z}$ and suppose that there exist $\lambda, \mu \in \mathbb{Z}$ such that $\lambda m + \mu n = 1$. Prove that m and n are relatively prime.
15. Suppose that $a, b \in \mathbb{Z}$ and $\gcd(a, b) = 1$. Prove that $\gcd(a^m, b^n) = 1$ for $m, n \in \mathbb{N}$.
16. Let $m, n \in \mathbb{N}$ and let $S = \{xm + yn \mid x, y \in \mathbb{Z}\} \subseteq \mathbb{Z}$. Prove that
    (i) $q \in \mathbb{Z}$ and $s, t \in S \Rightarrow qs \in S$ and $s + t \in S$.
    (ii) Assume that $S \neq \{0\}$. Use (i) to prove that $S = \{ad \mid a \in \mathbb{Z}\}$, where d is the first element > 0 in $S \cap \mathbb{N}$.
    (iii) Prove that $d = \gcd(m, n)$ (again assuming that $S \neq \{0\}$).
    This gives another proof of Lemma 1.5.7.
17. What is the smallest odd natural number that leaves a remainder of 2 when divided by 3 and a remainder of 3 when divided by 5?
18. Solve the system ([11][18])

$$X \equiv 17 \pmod{504},$$
$$X \equiv -4 \pmod{35},$$
$$X \equiv 33 \pmod{16},$$

of congruences in X.

19. Why does the following number game work?

   Ask anyone to select a number less than 60. Request him to perform the following operations. (i) Divide it by 3 and mention the remainder; suppose it to be a. (ii) Divide it by 4, and mention the remainder; suppose it to be b. (iii) Divide it by 5 and mention the remainder; suppose it to be c. Then the number selected is the remainder obtained by dividing $40a + 45b + 36c$ by 60.

20. (Quoted from [18])   An old woman goes to market and a horse steps on her basket and crushes her eggs. The rider offers to pay for the damages and asks her how many eggs she had brought. She does not remember the exact number, but when she had taken them out two at a time, there was one egg left. The same had happened when she picked them out three, four, five and six at a time, but when she took them out seven at a time they came out even (no eggs left). What is the smallest number of eggs she could have had?

21. On a desert island, five men and a monkey gather coconuts all day, then they go to sleep. The first man wakes up and takes his share. He divides the coconuts into five equal shares and gives the monkey the one coconut left over, hides his share and goes back to sleep. The second man wakes up, takes his fifth from the remaining pile; he too finds one extra and gives it to the monkey. Each of the remaining three men does likewise in turn. Find the minimum number of coconuts that must have been originally present.

22. Prove that $\varphi(n) = \varphi(2n)$ if n is odd.

23. It seems that $\varphi(n)$ is even when $n > 2$. Can you prove this without using the formula in subsection 1.8.3?

24. Suppose that $p_1, \ldots, p_N$ are the first N prime numbers. Is $p_1 \cdots \cdots p_N + 1$ a prime number? (hint: $2 \cdot 3 \cdot 5 \cdot 7 \cdot 11 \cdot 13 \cdot 17 \equiv -1$ (mod 19)).

25. Prove that n has to be a prime number if the Mersenne number $M_n = 2^n - 1$ is a prime number. Is M_n a prime number if n is a prime number?

26. Prove that if $2^n + 1$ is a prime then n is a power of 2 (hint: if $n = ab$, where b is odd, then $2^a + 1$ divides $2^n + 1$). The nth Fermat number F_n is defined as $2^{2^n} + 1$. Prove that F_0, F_1, F_2, F_3, F_4 are prime numbers.

27. Prove that F_m and F_n (see the previous exercise) are relatively prime if $m \neq n$ (hint: prove and use that $\prod_{i=0}^{n-1} F_i = F_n - 2$). Use this to prove that there are infinitely many prime numbers.

28. Find a prime factorization of 2419 in less than 3 minutes.

29.   (i) Let $p > 3$ be a prime number. Prove that for every a, $1 < a < p - 1$, there is a unique $b \neq a$, $1 < b < p - 1$, such that $ab \equiv 1$ (mod p).

(ii) Let p be a prime number. Prove that $(p-1)! \equiv -1 (\text{mod } p)$ (hint: think in pairs and apply (i)).

(iii) Suppose that $(n-1)! \equiv -1 (\text{mod } n)$, where $n \geq 2$. Is n a prime number?

The result in (ii) is called Wilson's theorem.

30. (i) Let p be a prime number. Prove that

$$p \mid \binom{p}{i} \qquad \text{for } 1 \leq i \leq p-1$$

using Lemma 1.8.3.

(ii) Prove that

$$(a+b)^p \equiv a^p + b^p (\text{mod } p)$$

for integers a, b and a prime number p (hint: use (i) or Corollary 1.9.2).

(iii) Suppose that

$$n \mid \binom{n}{i} \qquad \text{for } 1 \leq i \leq n-1.$$

Is n a prime number?

31. Prove unique factorization using that $\mathbb{N}$ is well ordered, by assuming that

$$M = \{n \in \mathbb{N} \setminus \{0\} \mid n \text{ does not have a unique factorization}\}$$

is a non-empty subset of $\mathbb{N}$. Let m denote the first element in M. Consider two different prime factorizations

$$m = p_1 \cdots p_r,$$
$$m = q_1 \cdots q_s$$

of m.

(i) Prove that

$$\{p_1, \ldots, p_r\} \cap \{q_1, \ldots, q_s\} = \emptyset.$$

(ii) Assume that $p_1 < q_1$. Use the fact that the number

$$n = p_1 \cdots p_r - p_1 q_2 \cdots q_s$$
$$= (q_1 - p_1) q_2 \cdots q_s$$

has a unique factorization to reach a contradiction.

This proof of unique factorization is from the classic text by Courant and Robbins [4].

32. What is the product of the greatest common divisor and the least common multiple?

33. Let $n \in \mathbb{N} \setminus \{0\}$ have the prime factorization

$$n = p_1^{e_1} \cdots p_m^{e_m},$$

where $p_i \neq p_j$ for $i \neq j$. Let $d(n) = |\mathrm{div}(n)|$ and

$$\sigma(n) = \sum_{d \in \mathrm{div}(n)} d$$

be respectively the number of natural divisors in n and the sum of the natural divisors in n.
   (i) Prove that $d(n) = (e_1 + 1) \cdots (e_m + 1)$.
   (ii) Prove that

$$\sigma(n) = \frac{p_1^{e_1+1} - 1}{p_1 - 1} \cdots \frac{p_m^{e_m+1} - 1}{p_m - 1}.$$

34. A number $n \in \mathbb{N}$ is called perfect if $\sigma(n) = 2n$. So, a number is perfect if it is the sum of its natural divisors except itself. Prove that if $2^{n+1} - 1$ is a prime number then $2^n(2^{n+1} - 1)$ is perfect.

35. Use GIMPS and a computer to find a perfect number with more than one million digits.

36. Let $n = p_1^{s_1} p_2^{s_2}$, where $p_1 \neq p_2$ are prime numbers. Prove that $\varphi(n) = (p_1^{s_1} - p_1^{s_1-1})(p_2^{s_2} - p_2^{s_2-1})$ by counting explicitly the number of natural numbers less than n that are relatively prime to n. If you like counting and combinatorics you may generalize this to give a proof of the formula for computing φ in subsection 1.8.3.

37. Prove that the fifth Fermat number (see Exercise 1.26) $F_5 = 2^{32} + 1$ is composite (this was first proved by Euler in 1739, thereby demolishing the conjecture that every F_n is prime) by using the following hints: $5^4 + 2^4 = 1 + 2^7 \cdot 5$ and

$$F_5 = (5^4 + 2^4)(2^7)^4 - 5^4(2^7)^4 + 1.$$

It is not known whether there is a Fermat number F_n that is prime for $n > 4$.

38. Suppose that $N = pq$ is the product of two different prime numbers p and q. Show that p and q are solutions to the equation

$$X^2 + (\varphi(N) - N - 1)X + N = 0.$$

This shows that (given $N = pq$) finding p and q is just as "difficult" as finding $\varphi(N)$.

39. **(HOF)** Around 1994 the following email circulated (partially quoted):

We are happy to announce that

RSA-129 = 114381625757888867669235779976146612010218296721242362562561 8429 \
35706935245733897830597123563958705058989075147599290026879543541
= 3490529510847650949147849619903898133417764638493387843990820577 *
32769132993266709549961988190834461413177642967992942539798288533

The encoded message published was

9686961375462206147714092225435588290575999112457431987469512093081 62 \
98225145708356931476622883989628013391990551829945157815154

This number came from an RSA encryption of the 'secret' message using the
public exponent 9007.

The symbol \ indicates that the number is continued on the next line. This
email announced that the original 1977 RSA challenge from Martin
Gardner's Scientific American column had been factored. It also gives the
encoded message using the following encoding: space $= 00$, A $= 01$, B $=
02, \ldots$. What was the secret message encrypted in 1977?

The factorization of RSA-129 was a real challenge, involving participants
in every corner of the world:

To find the factorization of RSA-129, we used the double large prime variation of
the multiple polynomial quadratic sieve factoring method. The sieving step took
approximately 5000 mips years, and was carried out in 8 months by about 600
volunteers from more than 20 countries, on all continents except Antarctica.
Combining the partial relations produced a sparse matrix of 569466 rows and
524338 columns. This matrix was reduced to a dense matrix of 188614 rows and
188160 columns using structured Gaussian elimination. Ordinary Gaussian
elimination on this matrix, consisting of 35489610240 bits (4.13 gigabyte), took
45 hours on a 16K MasPar MP-1 massively parallel computer. The first three
dependencies all turned out to be 'unlucky' and produced the trivial factor
RSA-129. The fourth dependency produced the above factorization.

We would like to thank everyone who contributed their time and effort to this
project. Without your help this would not have been possible.

Derek Atkins
Michael Graff
Arjen Lenstra
Paul Leyland

40. **(HOF)** Suppose that you are given e, d and N in the context of the RSA
    cryptosystem. The purpose of this exercise is to show that one can deduce
    the prime factorization $N = pq$ from this.
    (i) Show that the congruence $x^2 \equiv 1 \pmod{N}$ has four solutions
        modulo N (there are two more apart from the obvious $x = \pm 1$).
    (ii) Show that one of these solutions x satisfies $x \equiv -1 \pmod{p}$ and
        $x \equiv 1 \pmod{q}$. How can this be used to find p effectively?

(iii) Using that $\varphi(N)$ is even, deduce an effective probabilistic algorithm for finding p and q given e, d and N (you already know that $\varphi(N) \mid ed - 1$ and that $a^{\varphi(N)} \equiv 1 \,(\text{mod } N)$ for $\gcd(a, N) = 1$).

(iv) Why is it not secure to use the same N for different people in the RSA system?

41. Prove that $a^{N-1} \not\equiv 1 \,(\text{mod } N)$ if $\gcd(a, N) > 1$, where $a, N \in \mathbb{Z}$ and $N \geq 1$.

42. Prove that 899 is composite using only Corollary 1.9.2.

43. Prove that 15 is not a strong pseudoprime relative to 11.

44. Prove that 25 is a strong pseudoprime relative to 7.

45. Let $n = p_1 \cdots p_r$ be a product of primes, where $p_i \neq p_j$, $1 \leq i < j \leq r$. Suppose that $p_i - 1 \mid n - 1$ for $i = 1, \ldots, r$.
   (i) Prove that $a^{n-1} \equiv 1 \,(\text{mod } n)$ if $\gcd(a, n) = 1$.
   (ii) Prove that 561 is a Carmichael number.
   (iii) Give an example of a Carmichael number $\neq 561$.

46. Use Pollard's ρ-algorithm to factor $N = 10403$.

47. **(HOF)** Implement Pollard's ρ-algorithm using a computer language with infinite-precision integer arithmetic. Use the polynomial $f(X) = X^{2048} + 1$ and $X_0 = Y_0 = 3$ instead of $f(X) = X^2 + 1$ and $X_0 = 0$ to factor the eighth Fermat number

$$F_8 = 2^{2^8} + 1.$$

This is a number with 78 digits.

48. Use Pollard's $(p - 1)$-algorithm to factor $N = 295927$.

49. Part (ii) of this exercise uses linear algebra over the finite field $\mathbb{F}_2$ with two elements, which is detailed later in the book (see Chapter 3 and Appendix B).
   (i) Let $x = p_1^{m_1} \cdots p_r^{m_r}$ be a prime factorization of a positive natural number. Prove that x is a square if and only if all the exponents $m_1, \ldots, m_r$ are even.
   (ii) Suppose that the prime factorizations of $a_1, \ldots, a_n$ over the factor basis $P = \{p_1, \ldots, p_r\}$ (assume that all the a factor completely using primes from P) are

$$a_1 = p_1^{m_{11}} \cdots p_r^{m_{1r}},$$
$$a_2 = p_1^{m_{21}} \cdots p_r^{m_{2r}},$$
$$\vdots$$
$$a_n = p_1^{m_{n1}} \cdots p_r^{m_{nr}}.$$

Translate the problem of finding a subset $\{i_1, \ldots, i_s\}$ of $\{1, 2, \ldots, n\}$ such that $a_{i_1} \cdots a_{i_s}$ is a square into linear algebra over $\mathbb{F}_2$.

50. Let $f(X) = a_n X^n + \cdots + a_1 X + a_0$, where $a_i \in \mathbb{Z}$, $n \in \mathbb{N}$ and X is a variable. The degree of f is said to be n modulo $N \in \mathbb{Z}$ if $N \nmid a_n$.
    (i) Show that $X - a \mid X^n - a^n$, where $X, a \in \mathbb{Z}$ and $n \in \mathbb{N}$.
    (ii) Let $a, N \in \mathbb{Z}$. Show that if f has degree n modulo N and $f(a) \equiv 0$ (mod N) then $f(X) \equiv (X - a)g(X) \pmod{N}$, where g has degree $n - 1$ modulo N (use (i) and $f(X) \equiv f(X) - f(a) \pmod{N}$).
    (iii) Show that the congruence $f(X) \equiv 0 \pmod{p}$ has at most n incongruent solutions modulo p, if p is a prime and f has degree n modulo p. What if p is not a prime?

51. Let p be an odd prime.
    (i) Prove that the product of two quadratic residues modulo p is a quadratic residue modulo p.
    (ii) Prove that the product of two quadratic non-residues modulo p is a quadratic residue modulo p.
    (iii) Prove that the product of a quadratic residue modulo p and a quadratic non-residue modulo p is a quadratic non-residue modulo p.

52. Determine the quadratic residues and non-residues modulo 13.

53. Show that 3 is a quadratic residue modulo the prime p if $p \equiv 1 \pmod{12}$.

54. Compute

$$\left(\frac{7}{17} \right).$$

2 Groups

The concept of a group was first formalized by Cayley (1821–95) around 1854, but many mathematicians computed with group-like structures before that. In fact one of the main results in introductory group theory (see Theorem 2.2.8 below) was already known to Lagrange (1736–1813) in 1771. At this point we need to introduce groups in order to have a language that makes life easier. Dealing with numbers, we have encountered group-like structures several times already. By introducing the basic notions of group theory we get very simple (and nice) proofs of Euler's and Fermat's theorems on congruences (Theorem 1.7.2 and Corollary 1.9.2). By some mystery you are able to do much more powerful mathematics by introducing the three simple axioms defining a group. One point is worth singling out in this chapter: you will increase your level of abstraction from computing with elements in a set to computing with subsets of a set. In fact group theory puts the theory of congruences in a natural context and it will make sense to add and multiply subsets of $\mathbb{Z}$ consisting of numbers with the same remainder with respect to a positive integer. Groups are also useful outside the world of numbers. Using symmetric and alternating groups we will give a complete treatment of the 15-puzzle invented by Sam Loyd in 1878. Loyd offered a 1000-dollar prize for a correct solution. You can understand why this puzzle usually drives people nuts by reading subsection 2.9.5.

At the end of the chapter we treat actions of groups on sets. This is an extremely useful notion. We will apply actions of groups to combinatorics and counting and in the proof of the celebrated Sylow theorems.

2.1 Definition

A *composition* on a set G is a map $\circ : G \times G \to G$. The composition $\circ(g, h)$ is often written $g \circ h$ or gh.

Definition 2.1.1   A pair $(G, \circ)$ consisting of a set G and a composition $\circ :$ $G \times G \to G$ is called a *group* if it satisfies the following three properties.

(i) The composition is *associative*:

$$s_1 \circ (s_2 \circ s_3) = (s_1 \circ s_2) \circ s_3$$

for every $s_1, s_2, s_3 \in G$.

(ii) There is a *neutral* element $e \in G$ such that

$$e \circ s = s \qquad \text{and} \qquad s \circ e = s$$

for every $s \in G$.

(iii) For every $s \in G$ there is an *inverse* element $t \in G$ such that

$$s \circ t = e \qquad \text{and} \qquad t \circ s = e.$$

A group G is called *abelian* if $x \circ y = y \circ x$ for every $x, y \in G$. The number of elements $|G|$ in G is called the *order* of G.

The first few examples of groups arise in the world of numbers. The set of natural numbers $(\mathbb{N}, +)$ with the composition $+$ is not a group, since the neutral element would have to be 0, but then 1, for example, would not have an inverse element (there would not exist $x \in \mathbb{N}$ such that $x + 1 = 0$). This defect is repaired by introducing the set of integers $\mathbb{Z}$, which is an abelian group with the composition $+$. The rational numbers $(\mathbb{Q}, +)$ and the real numbers $(\mathbb{R}, +)$ are also abelian groups. The sets of non-zero rational numbers $(\mathbb{Q} \setminus \{0\}, \cdot)$ and non-zero real numbers $(\mathbb{R} \setminus \{0\}, \cdot)$ are abelian groups with multiplication as composition.

The axioms defining a group resemble the rules of chess. You can learn them in a few minutes. To become a skilled player, however, you need to see lots of examples of groups in many contexts. You have little or no insight in the concept of a group by just knowing (i)–(iii) above. The first question to ask is, why do we introduce this abstraction? To begin with let us see how congruences fit into this framework.

2.1.1   Groups and congruences

A group is a vast generalization of the integers $\mathbb{Z}$ with $+$. The advantage of working with $\mathbb{Z}$ instead of $\mathbb{N}$ is that every number $x \in \mathbb{Z}$ has an inverse $y = -x$, so that $x + y = 0$. In this context 0 is a neutral element for $+$, in that $x + 0 = x$ for every $x \in \mathbb{Z}$. One very important property is associativity, as mentioned above. This concept arises as an attempt to give meaning to the expression $x + y + z$, where $x, y, z \in \mathbb{Z}$. This expression only makes sense when we insert parentheses, to give $(x + y) + z$ or $x + (y + z)$, because $+$ is a map $\mathbb{Z} \times \mathbb{Z} \to \mathbb{Z}$. Associativity states that $(x + y) + z = x + (y + z)$ – it does not matter how you insert the parentheses.

Granting that $\mathbb{Z}$ is a group with the composition $+$, let us see how to build some new groups tied up with congruence modulo an integer. We will define addition, $+$, on subsets of $\mathbb{Z}$ given by $a + n\mathbb{Z} = \{a + nx \mid x \in \mathbb{Z}\}$, where $a, n \in \mathbb{Z}$. As an example we have

$$2 + 5\mathbb{Z} = \{\ldots, -8, -3, 2, 7, \ldots\}.$$

When working with numbers it was easy to spot when two elements were identical. Now, we will work with subsets. Two subsets are identical when they contain the same elements. You may check for example that $5 + 7\mathbb{Z} = 19 + 7\mathbb{Z}$. This is a special case of the following proposition.

Proposition 2.1.2 *Let $a, b, c \in \mathbb{Z}$. Then $a + c\mathbb{Z} = b + c\mathbb{Z}$ if and only if $a \equiv b \,(\mathrm{mod}\ c)$. Also $(a + c\mathbb{Z}) \cap (b + c\mathbb{Z}) = \emptyset$ if and only if $a \not\equiv b \,(\mathrm{mod}\ c)$.*

Proof. If $m \in a + c\mathbb{Z}$ then $m = a + cx$, where $x \in \mathbb{Z}$. If $a + c\mathbb{Z} = b + c\mathbb{Z}$ then $m \in b + c\mathbb{Z}$. This shows that $m = a + cx = b + cy$ for $y \in \mathbb{Z}$. Therefore $a - b = c(y - x)$ and $a \equiv b \,(\mathrm{mod}\ c)$. However, if $a \equiv b \,(\mathrm{mod}\ c)$ then $a = b + cx$ for $x \in \mathbb{Z}$. Therefore $a + c\mathbb{Z} = b + cx + c\mathbb{Z} = b + c\mathbb{Z}$, since $cx + c\mathbb{Z} = c\mathbb{Z}$. If $(a + c\mathbb{Z}) \cap (b + c\mathbb{Z}) \neq \emptyset$, then we may find $m, x, y \in \mathbb{Z}$ such that $m = a + cx = b + cy$. This gives $a - b = c(y - x)$ and therefore $a \equiv b \,(\mathrm{mod}\ c)$. This proves that if $(a + c\mathbb{Z}) \cap (b + c\mathbb{Z}) \neq \emptyset$ then $a + c\mathbb{Z} = b + c\mathbb{Z}$. $\square$

If $c > 0$ we have $a + c\mathbb{Z} = b + c\mathbb{Z}$ if and only if $[a]_c = [b]_c$, by Proposition 1.3.2(i). In this context we let $[x]$ denote the subset $x + c\mathbb{Z}$. Then $[x] = [[x]_c]$, so there can be only finitely many different subsets of the form $[x]$. These are given by the remainders $[0], [1], \ldots, [c-1]$ after division by c. Denote the set of these subsets by $\mathbb{Z}/c\mathbb{Z}$.

Example 2.1.3 Let $c = 3$. Then $\mathbb{Z}/3\mathbb{Z} = \{[0], [1], [2]\}$, where

$$[0] = \{\ldots, -6, -3, 0, 3, 6, \ldots\},$$
$$[1] = \{\ldots, -5, -2, 1, 4, 7, \ldots\},$$
$$[2] = \{\ldots, -4, -1, 2, 5, 8, \ldots\}.$$

Armed with these definitions we can add subsets $[x], [y] \in \mathbb{Z}/c\mathbb{Z}$ simply by defining $[x] + [y] = [x + y]$.

Notice the problem here! We need to check that if $[x] = [x']$ and $[y] = [y']$ then $[x + y] = [x' + y']$. But we have already done that in Proposition 1.3.4(i), where we proved that $x \equiv x' \pmod{c}$ and $y \equiv y' \pmod{c}$ implies that $x + y \equiv x' + y' \pmod{c}$. So by Proposition 2.1.2, the composition $+$ is well defined. Make sure you understand that there really is something to be checked here.

With the composition $+$ constructed in this way, $(\mathbb{Z}/c\mathbb{Z}, +)$ is a group of order c. The neutral element is the subset $[0] = c\mathbb{Z}$. The inverse element of $[x]$ is $[-x]$, and associativity holds because

$$([x] + [y]) + [z] = [x + y] + [z] = [(x + y) + z] = [x + (y + z)]$$
$$= [x] + [y + z] = [x] + ([y] + [z])$$

for $[x], [y], [z] \in \mathbb{Z}/c\mathbb{Z}$. Here we have used the fact that associativity holds in $(\mathbb{Z}, +)$. The group $(\mathbb{Z}/c\mathbb{Z}, +)$ is abelian since $[x] + [y] = [x + y] = [y + x] = [y] + [x]$ for every $[x], [y] \in \mathbb{Z}/c\mathbb{Z}$. If $c = 0$ then $x + c\mathbb{Z} = \{x\}$ and we simply recover $(\mathbb{Z}, +)$ as the group $(\mathbb{Z}/0\mathbb{Z}, +)$.

2.1.2 The composition table

Definition 2.1.4 When dealing with a finite group $(\{e, g_1, \ldots, g_r\}, \circ)$, the composition $\circ$ is often displayed in a *composition table*:

$\circ$	e	g_1	$\cdots$	g_j	$\cdots$	g_r
e	e	g_1	$\cdots$	g_j	$\cdots$	g_r
g_1	g_1	$g_1 \circ g_1$	$\cdots$	$g_1 \circ g_j$	$\cdots$	$g_1 \circ g_r$
$\vdots$	$\vdots$	$\vdots$	$\ddots$	$\vdots$	$\ddots$	$\vdots$
g_i	g_i	$g_i \circ g_1$	$\cdots$	$g_i \circ g_j$	$\cdots$	$g_i \circ g_r$
$\vdots$	$\vdots$	$\vdots$	$\ddots$	$\vdots$	$\ddots$	$\vdots$
g_r	g_r	$g_r \circ g_1$	$\cdots$	$g_r \circ g_j$	$\cdots$	$g_r \circ g_r$

Example 2.1.5 The composition table for the finite group $(\mathbb{Z}/4\mathbb{Z}, +)$ with elements $[0], [1], [2], [3]$ is

$+$	$[0]$	$[1]$	$[2]$	$[3]$
$[0]$	$[0]$	$[1]$	$[2]$	$[3]$
$[1]$	$[1]$	$[2]$	$[3]$	$[0]$
$[2]$	$[2]$	$[3]$	$[0]$	$[1]$
$[3]$	$[3]$	$[0]$	$[1]$	$[2]$

2.1.3 Associativity

Suppose that S is a set with a composition $S \times S \to S$, where (x, y) maps to xy. Assume that $x(yz) = (xy)z$ for every x, y, $z \in S$ (the composition is associative). Writing an expression like $s_1 s_2 s_3$ for s_1, s_2, $s_3 \in S$ is clearly nonsense, since the composition is only defined given two elements from S. We can make sense of it by (1) first evaluating $s_1 s_2$ and then composing with s_3 or (2) first evaluating $s_2 s_3$ and then composing with s_1 (from the left). Associativity says that these two ways of evaluating give the same result. Similarly, for four elements s_1, s_2, s_3, s_4 of a group, we have five ways of evaluating $s_1 s_2 s_3 s_4$:

$$s_1(s_2(s_3 s_4)),$$
$$s_1((s_2 s_3)s_4),$$
$$(s_1(s_2 s_3))s_4,$$
$$((s_1 s_2)s_3)s_4,$$
$$(s_1 s_2)(s_3 s_4).$$

You can use associativity to prove that these five different ways of evaluating $s_1 s_2 s_3 s_4$ all give the same result. There are 4862 ways[1] of evaluating the product $s_1 s_2 \cdots s_{10}$ of 10 elements. Associativity still proves that these are all the same. One can prove, using associativity, that any two ways of evaluating a product $s_1 s_2 \cdots s_n$ lead to the same result.

In general it is difficult to decide whether a composition on a set is associative. There is one exceedingly important case for which we have an associative composition. This is the case where S is the set of maps from a set X to itself and the composition is the usual composition of maps, in which fg is defined by $(fg)(x) = f(g(x))$ for f, $g \in S$ and $x \in X$. In this case $f(gh) = (fg)h$, since $(f(gh))(x)$ and $((fg)h)(x)$ are identical for every $x \in X$:

$$(f(gh))(x) = f((gh)(x)) = f(g(h(x))),$$
$$((fg)h)(x) = (fg)(h(x)) = f(g(h(x))).$$

2.1.4 The first non-abelian group

To show the wide application of groups, we give an example of a non-abelian group with six elements. You should keep referring to this example

---

[1] Computing the number of ways C_{n-1} of evaluating the product of n elements $s_1 s_2 \cdots s_n$ by inserting parentheses is a classical problem of combinatorics, referred to as Catalan's problem. One may prove that

$$C_n = \frac{1}{n+1}\binom{2n}{n}.$$

when new concepts are introduced. It contains all the ingredients of a good understanding.

Example 2.1.6 Let $X = \{1, 2, 3\}$ be a set consisting of three elements. Let G be the set of all bijective maps $X \to X$. Then G is a group with the usual composition of maps as composition (see the previous subsection). The neutral element e is the identity map $X \to X$. The element inverse to a given map $f : X \to X$ is the inverse map $f^{-1} : X \to X$, and the composition of maps is associative (we saw this in subsection 2.1.3). We can list the elements of G as follows:

$$e = \begin{pmatrix} 1 & 2 & 3 \\ 1 & 2 & 3 \end{pmatrix}, \qquad a = \begin{pmatrix} 1 & 2 & 3 \\ 2 & 1 & 3 \end{pmatrix}, \qquad b = \begin{pmatrix} 1 & 2 & 3 \\ 1 & 3 & 2 \end{pmatrix},$$

$$c = \begin{pmatrix} 1 & 2 & 3 \\ 3 & 2 & 1 \end{pmatrix}, \qquad d = \begin{pmatrix} 1 & 2 & 3 \\ 3 & 1 & 2 \end{pmatrix}, \qquad f = \begin{pmatrix} 1 & 2 & 3 \\ 2 & 3 & 1 \end{pmatrix}.$$

where for example $c : X \to X$ is the bijective map given by $c(1) = 3$, $c(2) = 2$, $c(3) = 1$. To compute ab you simply find what the map $a \circ b$ does to $1, 2, 3$. Now $ab(1) = a(b(1)) = 2$, $ab(2) = a(b(2)) = 3$ and $ab(3) = a(b(3)) = 1$. This shows that $ab = f$. The composition table (see subsection 2.1.2) is constructed using this reasoning.

$\circ$	e	a	b	c	d	f
e	e	a	b	c	d	f
a	a	e	f	d	c	b
b	b	d	e	f	a	c
c	c	f	d	e	b	a
d	d	b	c	a	f	e
f	f	c	a	b	e	d

The group G is also known as the symmetric group S_3. It is non-abelian since $ab \neq ba$.

2.1.5 Uniqueness of neutral and inverse elements

There can be only one neutral element in a group G. If $e' \in G$ were another then $e = e'e = e'$ by Definition 2.1.1(ii). Also, to every $g \in G$ there can be only one inverse element, h. Suppose that h' is an element satisfying $gh' = e$. Then $e = hg$ implies that $h' = (hg)h' = h(gh') = he = h$ by Definition 2.1.1 (iii).

Definition 2.1.7 Let $g \in G$ be an element of a group. Then we let $g^{-1} \in G$ denote the unique inverse element of g.

Example 2.1.8 Finding the inverse element of a product ab in a group is similar to inverting a product of invertible matrices. In fact,

$$(ab)(b^{-1}a^{-1}) = a(b(b^{-1}a^{-1})) = a(ea^{-1}) = aa^{-1} = e$$

shows that $(ab)^{-1} = b^{-1}a^{-1}$. The computation for $(b^{-1}a^{-1})(ab)$ is similar.

2.1.6 Multiplication by $g \in G$ is bijective

Suppose that G is a group and $g \in G$. Then there is a map $\varphi : G \to G$ given by $\varphi(x) = gx$. This map is bijective. We can prove this by giving the inverse map $\psi : G \to G$ to φ. Consider the map $\psi(x) = g^{-1}x$ from G to G. Then $\psi(\varphi(x)) = g^{-1}(gx) = (g^{-1}g)x = ex = x$ and $\varphi(\psi(x)) = g(g^{-1}x) = (gg^{-1})x = ex = x$. This proves that ψ is the inverse map of φ and therefore that φ is a bijection. Notice how all the properties of the group composition in Section 2.1.1 come into play.

In the same way one can prove that the map $\xi : G \to G$ given by $\xi(x) = xg$ is a bijection (see Exercise 2.1).

Example 2.1.9 What does a group G of order three look like? There must be a (unique) neutral element $e \in G$ and two other elements $a, b \in G$. To describe the composition $\circ : G \times G \to G$ we fill out the composition table:

$\circ$	e	a	b
e	$e \circ e$	$e \circ a$	$e \circ b$
a	$a \circ e$	$a \circ a$	$a \circ b$
b	$b \circ e$	$b \circ a$	$b \circ b$

We know that $e \circ a = a \circ e = a$ and $e \circ b = b \circ e = b$. This gives us the partial table

$\circ$	e	a	b
e	e	a	b
a	a		
b	b		

An important point is that an element in a group can only occur once in a row (or a column) of the composition table. The reason is that multiplication by a group element is bijective. Using this fact, there is only one way to complete the table:

$\circ$	e	a	b
e	e	a	b
a	a	b	e
b	b	e	a

We have proved that there is only one way of filling out the composition table for a group of order three (the same holds for any prime number. We will prove this in Proposition 2.7.2).

2.1.7 More examples of groups

The only way to understand the concept of a group is to study its many incarnations. We give some more important examples in this subsection.

Example 2.1.10 Using matrices we will give an example of an infinite non-abelian group. Let

$$\mathrm{GL}_2(\mathbb{R}) = \left\{ \begin{pmatrix} a & b \\ c & d \end{pmatrix} \mid a, b, c, d \in \mathbb{R} \quad \text{and} \quad \det \begin{pmatrix} a & b \\ c & d \end{pmatrix} = ad - bc \neq 0 \right\}$$

denote the set of 2×2 matrices with real entries and non-zero determinant. The multiplication of matrices gives a composition on $\mathrm{GL}_2(\mathbb{R})$ as $\det(AB) = \det(A)\det(B)$ for $A, B \in \mathrm{GL}_2(\mathbb{R})$. One may check that it is associative by explicit computation (or identify matrix multiplication with the composition of linear maps). The identity matrix is the neutral element in $\mathrm{GL}_2(\mathbb{R})$ and the inverse matrix A^{-1} is the inverse element of $A \in \mathrm{GL}_2(\mathbb{R})$; recall that $\det(A^{-1}) = \det(A)^{-1}$. This group is called the general linear group, or more precisely the 2×2 general linear group. It is a non-abelian group (can you find $A, B \in \mathrm{GL}_2(\mathbb{R})$ such that $AB \neq BA$?).

Example 2.1.11 Recall that the transpose of a 2×2 matrix is given by

$$\begin{pmatrix} a & b \\ c & d \end{pmatrix}^t = \begin{pmatrix} a & c \\ b & d \end{pmatrix}.$$

A matrix $A \in \mathrm{GL}_2(\mathbb{R})$ is called orthogonal if $AA^t = I$, where I is the identity matrix. The set of 2×2 orthogonal matrices is denoted $O_2(\mathbb{R})$. Matrix multiplication is in fact a composition on $O_2(\mathbb{R})$. This follows from the identity $(AB)^t = B^t A^t$ for $A, B \in O_2(\mathbb{R})$. Since $(A^{-1})^t = (A^t)^{-1}$, $O_2(\mathbb{R})$ is a group with matrix multiplication as composition. It is called the orthogonal group (or more precisely the 2×2 orthogonal group).

Example 2.1.12 An isometry of the plane $\mathbb{R}^2$ is a map $\varphi : \mathbb{R}^2 \to \mathbb{R}^2$ preserving the Euclidean distance between any two points in $\mathbb{R}^2$:

$$|\varphi(x) - \varphi(y)| = |x - y|$$

for every $x, y \in \mathbb{R}^2$. One can prove that an isometry fixing the origin $(0, 0)$ is a linear invertible map (see Exercise 2.8). We call such an isometry linear. The set L of linear isometries of $\mathbb{R}^2$ is a group with respect to the usual composition of maps. Let us prove this in detail. First we need to see that the composition $\varphi_1 \circ \varphi_2$ of two linear isometries $\varphi_1, \varphi_2 \in L$ is again a linear isometry. This is to make sure that $\circ$ really is a composition on L. Given $x, y \in \mathbb{R}^2$,

$$|\varphi_1(\varphi_2(x)) - \varphi_1(\varphi_2(y))| = |\varphi_2(x) - \varphi_2(y)| = |x - y|.$$

This proves that $\varphi_1 \circ \varphi_2$ is an isometry. Since $\varphi_1(\varphi_2((0, 0))) = \varphi_1((0, 0)) = (0, 0)$, it must be linear. The neutral element in L with respect to $\circ$ is the identity map. Also, if $\varphi \in L$ then $\varphi^{-1} \in L$ (we know that φ is bijective): we need to prove that

$$|\varphi^{-1}(x) - \varphi^{-1}(y)| = |x - y|$$

for every $x, y \in \mathbb{R}^2$. But since φ is surjective we can find $x', y' \in \mathbb{R}^2$ such that $x = \varphi(x')$ and $y = \varphi(y')$. Therefore

$$|\varphi^{-1}(x) - \varphi^{-1}(y)| = |x' - y'| = |\varphi(x') - \varphi(y')| = |x - y|.$$

This shows that $\varphi^{-1} \in L$. We know from subsection 2.1.3 that the composition of maps is associative. Therefore $\circ$ is associative on L. In total we have proved that $(L, \circ)$ really is a group. But what are the maps in L? Let us compute the matrix of a linear isometry $\varphi : \mathbb{R}^2 \to \mathbb{R}^2$ in the standard basis $e_1 = (1, 0)$ and $e_2 = (0, 1)$ of $\mathbb{R}^2$.

We know that $|\varphi(e_1)| = |\varphi(e_1) - \varphi((0, 0))| = |e_1| = 1$. This implies that $\varphi(e_1) = (\cos(t), \sin(t))$ for $t \in \mathbb{R}$. However, since $|e_1 - e_2| = |(1, -1)| = \sqrt{2}$, we get $|(\cos(t), \sin(t)) - \varphi(e_2)| = \sqrt{2}$. This gives $\varphi(e_2) = (-\sin(t), \cos(t)) = v_1$ or $\varphi(e_2) = (\sin(t), -\cos(t)) = v_2$, as seen in the following diagram:

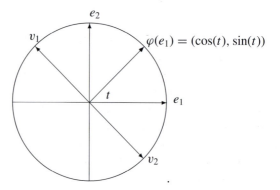

So, given a linear isometry φ we have two possibilities for its matrix when $\varphi(e_1)$ is determined by $\varphi(e_1) = (\cos(t), \sin(t))$. The first is represented by the matrix

$$\begin{pmatrix} \cos(t) & -\sin(t) \\ \sin(t) & \cos(t) \end{pmatrix};$$

this corresponds to a rotation. The second one is given by the matrix

$$\begin{pmatrix} \cos(t) & \sin(t) \\ \sin(t) & -\cos(t) \end{pmatrix};$$

this corresponds to a reflection in the line $L = \{(r\cos(t/2), r\sin(t/2)) \mid r \in \mathbb{R}\}$.

We have demystified the "complicated" term linear isometry and proved that we are dealing with rotations and reflections of the plane. Since L is a group the composition of a rotation and a reflection must be a reflection or a rotation. Which one is it?

If you prefer a more algebraic way of looking at the group L, you may prove that matrices of linear isometries are orthogonal (see Example 2.1.11). In fact, as we will see later, in Example 2.4.7, L is, in a specific sense, the same group as $O_2(\mathbb{R})$ from Example 2.1.11.

Example 2.1.13 Consider the subset $G \subset L$ of linear isometries (see Example 2.1.12) of $\mathbb{R}^2$ mapping an equilateral triangle K centered at $(0,0)$ to itself. Thus

$$G = \{\varphi \in L \mid \varphi(K) = K\}.$$

Let us check that G is a group with respect to the composition of maps. First (as in Example 2.1.12) we need to check that $\varphi_1 \circ \varphi_2 \in G$ when $\varphi_1, \varphi_2 \in G$. This is definitely true, since $\varphi_1(\varphi_2(K)) = \varphi_1(K) = K$ when $\varphi_1, \varphi_2 \in K$. The identity map is the neutral element in G. If $\varphi \in G$, we need to prove that $\varphi^{-1} \in G$. This also holds, since $\varphi^{-1}(K) = \varphi^{-1}(\varphi(K)) = K$. Again, we know from subsection 2.1.3 that composition of maps is associative. Therefore $\circ$ is an associative composition, just as in Example 2.1.12. We have proved that $(G, \circ)$ is a group. What are the maps in G? These are the rotations and reflections preserving the equilateral triangle.

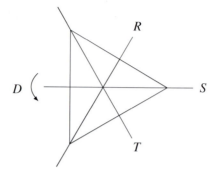

The only reflections preserving K are the reflections in the lines R, S and T above. The only rotations preserving K are I, D, D^2, where I is the identity map, D is a rotation of $2\pi/3$ (depicted above) and $E = D^2$ is a rotation of $4\pi/3$. Now it follows that

$$G = \{I, R, S, T, D, E\}$$

and that this finite subset of L really is a group. The composition table can be written down through explicit sketching:

$\circ$	I	R	S	T	D	E
I	I	R	S	T	D	E
R	R	I	D	E	S	T
S	S	E	I	D	T	R
T	T	D	E	I	R	S
D	D	T	R	S	E	I
E	E	S	T	R	I	D

Usually G is denoted D_3 and called the dihedral group of order 6. We will see later, in Example 2.4.6, that it is in a specific sense the same group as the group S_3 from Example 2.1.6 .

2.2 Subgroups and cosets

In Example 2.1.12 we saw an example of a group L containing a subset G that is a group with respect to the composition of L. Again in Example 2.1.11, the subset $O_2(\mathbb{R})$ of $GL_2(\mathbb{R})$ turned out to be a group with respect to the composition of $GL_2(\mathbb{R})$. This leads us to the concept of a subgroup.

Definition 2.2.1 A *subgroup* of a group G is a non-empty subset $H \subseteq G$ such that the composition of G makes H into a group, i.e. H is a subgroup of G if and only if

(i) $e \in H$,
(ii) $x^{-1} \in H$ for every $x \in H$,
(iii) $xy \in H$ for every $x, y \in H$.

If you revisit Example 2.1.13, you will see that we actually proved there that G is a subgroup of L, by verifying steps (i)–(iii) above.

Example 2.2.2 Returning to the group S_3 from Example 2.1.6, you can check that the two subsets $\{e, a\}$ and $\{e, f, d\}$ are subgroups by looking at the composition table for S_3.

2.2.1 Subgroups of $\mathbb{Z}$

We know that $(\mathbb{Z}, +)$ is a group. In the language of groups, division with remainder (Theorem 1.2.1) has a very pretty consequence.

Proposition 2.2.3 *Let H be a subgroup of $(\mathbb{Z}, +)$. Then*

$$H = d\mathbb{Z} = \{dn \mid n \in \mathbb{Z}\} = \{\ldots, -2d, -d, 0, d, 2d, \ldots\}$$

for a unique natural number $d \in \mathbb{N}$.

Proof. If $H = \{0\}$ we may put $d = 0$. Assume that $H \neq \{0\}$. Then $\mathbb{N} \cap H$ contains a smallest natural number $d > 0$ (why?). We claim that $H = d\mathbb{Z}$. It follows that $-d \in H$, since $d \in H$ and H is a subgroup. Again using that H is a subgroup, we get $-d + (-d) = -2d \in H$ and $d + d = 2d \in H$, $-2d + (-d) = -3d \in H$ and $2d + d = 3d, \ldots$. This shows that $nd \in H$ for every $n \in \mathbb{Z}$. Therefore $d\mathbb{Z} \subseteq H$.

Now let $m \in H$. Division with remainder gives $m = qd + r$, where $0 \leq r < d$. Since H is a subgroup, $m \in H$ and $d \in H$, we get $-qd \in H$ and $r = m - qd \in H$. But $r \geq 0$ is a natural number $< d$ in H. This means that $r = 0$, so that $m = qd$ and $H \subseteq d\mathbb{Z}$. Therefore $H = d\mathbb{Z}$. $\qquad\square$

2.2.2 Cosets

Let H be a subgroup of G and $g \in G$. Then the subset

$$gH = \{gh \mid h \in H\} \subseteq G$$

is called a *left coset* of H. Similarly we call the subset

$$Hg = \{hg \mid h \in H\} \subseteq G$$

a *right coset* of H. The set of left cosets of H is denoted G/H. The set of right cosets of H is denoted $H\backslash G$.

Example 2.2.4   If $G = (\mathbb{Z}, +)$ and $H = 3\mathbb{Z}$ then

$$\mathbb{Z}/3\mathbb{Z} = \{3\mathbb{Z}, 1 + 3\mathbb{Z}, 2 + 3\mathbb{Z}\}.$$

Notice that $1 + 3\mathbb{Z} = 4 + 3\mathbb{Z}$. This illustrates the fact that you can have different ways of representing the same left coset: if H is a subgroup of G and $g_1 H = g_2 H$ then g_1 and g_2 are not necessarily equal.

Example 2.2.5   Let H denote the subgroup $\{e, a\}$ of the group S_3 from Example 2.1.6. Let us list all the left and right cosets of H just using the definitions. First the left cosets:

$$\begin{aligned}
eH &= \{ee, ea\} &&= \{e, a\}, \\
aH &= \{ae, aa\} &&= \{a, e\}, \\
bH &= \{be, ba\} &&= \{b, d\}, \\
cH &= \{ce, ca\} &&= \{c, f\}, \\
dH &= \{de, da\} &&= \{d, b\}, \\
fH &= \{fe, fa\} &&= \{f, c\}.
\end{aligned}$$

We can already spot some interesting phenomena. It seems that left cosets are either equal or disjoint. Also, $eH = aH, bH = dH$ and $cH = fH$. This means that $G/H = \{H, bH, cH\}$. Let us carry out the same computations for the right cosets:

$$\begin{aligned}
He &= \{ee, ae\} &&= \{e, a\}, \\
Ha &= \{ea, aa\} &&= \{a, e\}, \\
Hb &= \{eb, ab\} &&= \{b, f\}, \\
Hc &= \{ec, ac\} &&= \{c, d\}, \\
Hd &= \{ed, ad\} &&= \{d, c\}, \\
Hf &= \{ef, af\} &&= \{f, b\}.
\end{aligned}$$

Here we have $He = Ha$, $Hb = Hf$ and $Hc = Hd$. This means that $H\backslash G = \{H, Hb, Hc\}$.

With this concrete example at hand, the following lemma should make sense, even though it might appear abstract at a first reading.

Lemma 2.2.6 *Let H be a subgroup of a group G and let $x, y \in G$. Then*

(i) $x \in xH$,
(ii) $xH = yH \iff x^{-1}y \in H$,
(iii) *If* $xH \neq yH$ *then* $xH \cap yH = \emptyset$,
(iv) *The map* $\varphi : H \to xH$ *given by* $\varphi(h) = xh$ *is bijective.*

Proof. Clearly $x \in xH$, since $x = xe$ and $e \in H$. This proves (i). If $xH = yH$ then $xh = ye = y$ for some $h \in H$. This implies that $x^{-1}y = h \in H$. If $x^{-1}y = h \in H$ then $y = xh$. Therefore $yH \subseteq xH$. Since $x = yh^{-1}$, we get $xH \subseteq yH$, so that $xH = yH$. This proves (ii). Suppose that $z \in xH \cap yH$. Then $z = xh_1 = yh_2$ for suitable $h_1, h_2 \in H$. But this shows that $x^{-1}y \in H$ and thus that $xH = yH$ by (ii). Therefore (iii) holds. Since φ is multiplication by x it follows by from subsection 2.1.6 that φ is bijective. It is simply the multiplication map restricted to the subgroup H. This proves (iv). $\square$

To connect with the theory of numbers, look at the subgroup $d\mathbb{Z}$ of $\mathbb{Z}$, where $d \in \mathbb{N}$. In this context Lemma 2.2.6(ii) says that $a + d\mathbb{Z} = b + d\mathbb{Z}$ if and only if $b - a \in d\mathbb{Z}$. Now $b - a \in d\mathbb{Z}$ means that $d \mid b - a$ or $a \equiv b \pmod{d}$ in the language of congruences. This is what we obtained in Proposition 2.1.2 without knowing about cosets.

Corollary 2.2.7 *Let H be a subgroup of G. Then*

$$G = \bigcup_{g \in G} gH,$$

and if $g_1 H \neq g_2 H$ *then* $g_1 H \cap g_2 H = \emptyset$.

Proof. Since $g \in gH$ (Lemma 2.2.6(i)) for every $g \in G$, we see that $G = \bigcup_{g \in G} gH$. If $g_1 H \neq g_2 H$ then $g_1 H \cap g_2 H = \emptyset$ by Lemma 2.2.6(iii). $\square$

We are now able to prove the Lagrange index theorem. Lagrange did not have the concept of an abstract group. He worked in the context of solutions to algebraic equations.

Theorem 2.2.8 (Lagrange) *If $H \subseteq G$ is a subgroup of a finite group G then*

$$|G| = |G/H||H|.$$

The order of a subgroup divides the order of the group.

Proof. Let gH be a coset in G/H. By Lemma 2.2.6(iv) there is a bijection between gH and H. This shows that gH has the same number of elements as H. Since G is the union of the cosets and different cosets are disjoint, by Corollary 2.2.7, the order of G must be the number of cosets times the order of H. This shows that $|G| = |G/H||H|$ and that $|H|$ divides $|G|$. $\square$

Definition 2.2.9   The number of cosets $|G/H|$ is called the *index* of H in G. It is denoted $[G : H]$.

Lagrange's theorem says that the order of a subgroup H divides the order of the group G. Suppose that d is a divisor in the order of a finite group G. Does G contain a subgroup of order d? After having digested Section 2.9 you will be able to solve Exercise 2.41, which answers this question negatively.

2.3 Normal subgroups

Let H be a subgroup of a group G. In a very important special case it is possible to make the set of left cosets, G/H, into a group inheriting the composition of G. What is the natural way of doing this? The set G/H consists of certain subsets of G called left cosets. We would like to compose two left cosets and get a new left coset. Why not compose subsets of G? Define

$$XY = \{xy \mid x \in X, y \in Y\}$$

for arbitrary subsets $X, Y \subseteq G$. This is a composition on the set of subsets of G, which is associative because the composition in G is associative. We would like this composition on subsets to give a composition on left cosets viewed as subsets. This is not necessarily so. Take a look back at Example 2.2.5. Here

$$(bH)(cH) = \{b, d\}\{c, f\} = \{bc, bf, dc, df\} = \{f, c, a, e\},$$

which is not a left coset. The key is the following.

Proposition 2.3.1   *Let H be a subgroup of a group G. If $gH = Hg$ for every $g \in G$ then*

$$(xH)(yH) = (xy)H$$

for every $x, y \in G$.

Proof. The inclusion $(xH)(yH) \supseteq (xy)H$ holds without any assumptions on H: if $(xy)h$ is an element of $(xy)H$ then $(xy)h = (xe)(yh) \in (xH)(yH)$. Let us show that $(xH)(yH) \subseteq (xy)H$. Let $(xh_1)(yh_2) \in (xH)(yH)$, where $h_1, h_2 \in H$. It follows that $(xh_1)(yh_2) = x((h_1y)h_2) = x((yh_3)h_2) = (xy)(h_3h_2)$ for a suitable $h_3 \in H$, since $Hy = yH$. This shows that $(xH)(yH) \subseteq (xy)H$. $\square$

This leads to the following definition.

Definition 2.3.2   A subgroup N of a group G is called *normal* if

$$gNg^{-1} = \{gng^{-1} \mid n \in N\} = N$$

for every $g \in G$.

A normal subgroup N of G satisfies $gN = Ng$ for every $g \in G$ (see Exercise 2.13).

Corollary 2.3.3   *Let N be a normal subgroup of the group G. Then the composition of subsets makes G/N into a group and*

$$(g_1 N)(g_2 N) = (g_1 g_2)N$$

for $g_1 N, g_2 N \in G/N$.

Proof.   We know that composition of subsets is associative and we have verified the above identity $(g_1N)(g_2N) = (g_1g_2)N$ for arbitrary $g_1, g_2 \in G$ in Proposition 2.3.1. So, the multiplication of subsets of G gives a composition on G/N (notice once more that it is crucial that N is normal). The neutral element is the left coset $eN = N$. The inverse element $(gN)^{-1}$ is $g^{-1}N$ for $gN \in G/N$. Therefore G/N is a group with this composition. $\square$

Definition 2.3.4   Let N be a normal subgroup of G. The group G/N is called a *quotient group.*

Example 2.3.5   The subset $H = \{e, a\} \subseteq S_3$ is a subgroup of S_3 (using the notation of Example 2.1.6). It is not normal, since the left coset $bH = \{b, ba\}$ is not equal to the right coset $Hb = \{b, ab\}$. This follows from the fact that $ab \neq ba$ as we have already seen. However, $K = \{e, d, f\}$ is a normal subgroup of S_3.

A subgroup of an abelian group is normal (see Exercise 2.14). Suppose that G is a group with the property that every subgroup in it is normal. Is G abelian? The answer is a somewhat surprising "no." The smallest non-abelian group for which every subgroup is normal is the quaternion group with eight elements (see Exercise 2.17).

Lemma 2.3.6 *Let H and K, where H is normal, be subgroups of a group G. Then HK is a subgroup of G.*

Proof. Clearly $e \in HK$. If $x \in H$, $y \in K$ then $(xy)^{-1} = (y^{-1}x^{-1}y)y^{-1} \in HK$. If furthermore $x' \in H$, $y' \in K$ then $(xy)(x'y') = (x(yx'y^{-1}))yy' \in HK$. $\square$

2.3.1 Quotient groups of the integers

Consider the subgroup $n\mathbb{Z}$ of $\mathbb{Z}$. This is a normal subgroup since $\mathbb{Z}$ is abelian. The quotient group $\mathbb{Z}/n\mathbb{Z}$ may appear abstract until you realize that it is exactly the same group as that defined at the start of subsection 2.1.1. The elements of $\mathbb{Z}/n\mathbb{Z}$ have the form $[x] = x + n\mathbb{Z}$, where $x \in \mathbb{Z}$. They are composed (here, added) using the familiar rule $[x] + [y] = [x + y]$. This is an application of Corollary 2.3.3.

The elements $[a] = a + n\mathbb{Z}$ of $\mathbb{Z}/n\mathbb{Z}$, where $a \in \mathbb{Z}$, are called *residue classes*. If $n > 0$ the residue classes of $\mathbb{Z}/n\mathbb{Z}$ are $\{[0], [1], \ldots, [n-1]\}$ – represented by the remainders after dividing by n.

2.3.2 The multiplicative group of prime residue classes

Looking at the set $\mathbb{Z}/n\mathbb{Z}$, where $n > 0$, can we multiply residue classes via $[a][b] = [ab]$ using ordinary multiplication in $\mathbb{Z}$ and get a group? We need to check that this makes sense. It may be possible that $[a] = [a']$ and $[b] = [b']$ but $[ab] \neq [a'b']$. This would make our definition meaningless. It would mean that $[a][b]$ has several values depending on the elements you choose in $[a]$ and $[b]$. Fortunately it does make sense since $[a] = [a']$ and $[b] = [b']$ can be rewritten as $a \equiv a' \pmod{d}$ and $b \equiv b' \pmod{n}$. Now Proposition 1.3.4 implies that $ab \equiv a'b' \pmod{n}$ or $[ab] = [a'b']$. So we get a well defined composition on $\mathbb{Z}/n\mathbb{Z}$. It is associative with neutral element $[1] = 1 + n\mathbb{Z}$, but not every element has an inverse. To begin with, $[a][0] = [0]$ for every $[a] \in n\mathbb{Z}$, so $[0]$ cannot have an inverse. Suppose we put $G = \mathbb{Z}/n\mathbb{Z} \setminus \{[0]\}$. This is still not good enough. Take the example $n = 4$; here $[2][2] = [0] \notin G$. The answer is to look at residue classes $[a] = a + n\mathbb{Z}$ with $\gcd(a, n) = 1$. You can easily check that if $a + n\mathbb{Z} = b + n\mathbb{Z}$ and $\gcd(a, n) = 1$ then $\gcd(b, n) = 1$. These residue classes are called *prime residue classes*. We let

$$(\mathbb{Z}/n\mathbb{Z})^* = \{[a] \in \mathbb{Z}/n\mathbb{Z} \mid \gcd(a, n) = 1\}$$

for $n \in \mathbb{N}$. The composition $[a][b] = [ab]$ is a composition on $(\mathbb{Z}/n\mathbb{Z})^*$, since $\gcd(a, n) = 1$ and $\gcd(b, n) = 1$ implies that $\gcd(ab, n) = 1$ (Corollary 1.5.11). Let $[a] \in (\mathbb{Z}/n\mathbb{Z})^*$. Then we can find $\lambda, \mu \in \mathbb{Z}$ such that $\lambda a + \mu n = 1$ by Lemma 1.5.7. In particular this gives $\gcd(\lambda, n) = 1$ and $[\lambda a] = [1]$, since $[\lambda a + \mu n] = [\lambda a] + [\mu n] = [\lambda a] + [0] = [\lambda a]$. But then $[\lambda]$ is the inverse element of $[a]$, since $[a][\lambda] = [a\lambda] = [1]$. We have proved that $(\mathbb{Z}/n\mathbb{Z})^*$ is a group with multiplication of residue classes as composition. The order of $(\mathbb{Z}/n\mathbb{Z})^*$ is $\varphi(n)$ for $n > 0$.

Example 2.3.7   Consider the group $(\mathbb{Z}/34\mathbb{Z})^*$. Then $[13] \in (\mathbb{Z}/34\mathbb{Z})^*$. In Example 1.5.3 we saw using the extended Euclidean algorithm that

$$5 \cdot 34 - 13 \cdot 13 = 1.$$

This implies that the inverse element of $[13]$ in $(\mathbb{Z}/34\mathbb{Z})^*$ is $[13]^{-1} = [21]$ (why?).

Example 2.3.8   If $n = 8$ then $(\mathbb{Z}/n\mathbb{Z})^*$ has the composition table

$\cdot$	$[1]$	$[3]$	$[5]$	$[7]$
$[1]$	$[1]$	$[3]$	$[5]$	$[7]$
$[3]$	$[3]$	$[1]$	$[7]$	$[5]$
$[5]$	$[5]$	$[7]$	$[1]$	$[3]$
$[7]$	$[7]$	$[5]$	$[3]$	$[1]$

The group $(\mathbb{Z}/n\mathbb{Z})^*$ is a much more subtle abelian group than $\mathbb{Z}/n\mathbb{Z}$. For one thing, the order of $(\mathbb{Z}/n\mathbb{Z})^*$ is $\varphi(n)$, a quantity difficult to compute as we have seen in Chapter 1. Later $(\mathbb{Z}/n\mathbb{Z})^*$ will appear more elegantly as the group of units in the ring $\mathbb{Z}/n\mathbb{Z}$.

The two groups $\mathbb{Z}/4\mathbb{Z}$ and $(\mathbb{Z}/8\mathbb{Z})^*$ both have four elements. They are abelian but quite different. In fact $G = (\mathbb{Z}/8\mathbb{Z})^*$ has the property that $g \cdot g = [1] = e$ for every $g \in G$. This is not shared by $\mathbb{Z}/4\mathbb{Z}$, where $[1] + [1] = [2] \neq [0] = e$. We require a tool to distinguish groups. We need to study maps between them that preserve their respective compositions.

2.4 Group homomorphisms

In what follows we will abuse notation somewhat by not writing the composition of elements explicitly. As before, we will write e for the neutral element in a group. It will be clear from the context to which group the composition and the neutral element refer.

Definition 2.4.1 Let G and K be groups. A map $f : G \to K$ is called a *group homomorphism* if $f(xy) = f(x)f(y)$ for every $x, y \in G$.

Example 2.4.2 The exponential function e^x is a group homomorphism from $(\mathbb{R}, +)$ to $(\mathbb{R}_{>0}, \cdot)$, where $\mathbb{R}_{>0} = \{x \in \mathbb{R} \mid x > 0\}$. This is the well known rule $e^{x+y} = e^x e^y$ for every $x, y \in \mathbb{R}$.

Example 2.4.3 The determinant

$$\det : \mathrm{GL}_2(\mathbb{R}) \to (\mathbb{R} \setminus \{0\}, \cdot)$$

is a group homomorphism (here $\cdot$ denotes multiplication). This is the well known rule $\det(AB) = \det(A)\det(B)$ for $A, B \in \mathrm{GL}_2(\mathbb{R})$.

Example 2.4.4 Let N be a normal subgroup of the group G. Then $\pi : G \to G/N$ given by $\pi(g) = gN$ is a group homomorphism. This follows from Corollary 2.3.3.

Definition 2.4.5 The kernel of a group homomorphism $f : G \to K$ is

$$\mathrm{Ker}\, f = \{g \in G \mid f(g) = e\}.$$

The image of f is $f(G) = \{f(g) \mid g \in G\} \subseteq K$. A bijective group homomorphism is called a *group isomorphism*. A group isomorphism $f : G \to K$ is denoted $f : G \xrightarrow{\sim} K$ and we write $G \cong K$ and say that G and K are *isomorphic*.

Isomorphisms between groups may appear a bit abstract at first. In the world of groups, isomorphic groups are considered as the same. For all practical purposes they have the same composition tables.

Example 2.4.6 Recall the groups S_3 (Example 2.1.6) and D_3 (Example 2.1.13). They are isomorphic. To prove this we give a map $f : D_3 \to S_3$ and prove that it is a group isomorphism. We number the corners in the equilateral triangle by 1, 2 and 3:

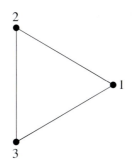

Given a rotation or a reflection $\sigma \in D_3$ it is easy to see that it must map a corner to a corner (if you do not believe this, you can go through the elements I, R, S, T, D, E of D_3 and check it). This enables us to construct a homomorphism φ from D_3 to S_3 given by

$$\varphi(\sigma) = \begin{pmatrix} 1 & 2 & 3 \\ \sigma(1) & \sigma(2) & \sigma(3) \end{pmatrix},$$

where $\sigma \in D_3$. Since σ is a bijective map $\mathbb{R}^2 \to \mathbb{R}^2$ is must also give a bijective map $\{1, 2, 3\} \to \{1, 2, 3\}$. Thus $\varphi(\sigma) \in S_3$. In order for φ to be a homomorphism, we must prove that $\varphi(\sigma_1 \circ \sigma_2) = \varphi(\sigma_1) \circ \varphi(\sigma_2)$. If you plug in the above form for $\varphi(\sigma)$, you will see that $\varphi(\sigma_1 \circ \sigma_2)(i) = (\varphi(\sigma_1) \circ \varphi(\sigma_2))(i) = \sigma_1(\sigma_2(i))$ for $i = 1, 2, 3$, so that φ really is a group homomorphism. Since a linear map $\mathbb{R}^2 \to \mathbb{R}^2$ is uniquely determined by its values on two linearly independent vectors, φ must be injective. So we have an injective group homomorphism $\varphi : D_3 \to S_3$. Since D_3 and S_3 are both of order 6, φ must be a group isomorphism.

Example 2.4.7 Let us prove that the groups L (Example 2.1.12) and $O_2(\mathbb{R})$ (Example 2.1.11) are isomorphic. There is a natural map $\varphi : L \to O_2(\mathbb{R})$. This is given simply by defining $\varphi(f)$ to be the matrix representing f in the natural basis e_1 and e_2. So if $f(e_1) = ae_1 + be_2$ and $f(e_2) = ce_1 + de_2$, where $a, b, c, d \in \mathbb{R}$, we put

$$\varphi(f) = \begin{pmatrix} a & b \\ c & d \end{pmatrix}.$$

From linear algebra it is known that $\varphi(f \circ g) = \varphi(f)\varphi(g)$ – composition of linear maps corresponds to multiplication of their matrices. So φ is a group homomorphism. From Example 2.1.12 you see that $\varphi(f) \in O_2(\mathbb{R})$. Given an orthogonal matrix

$$\begin{pmatrix} a & b \\ c & d \end{pmatrix}$$

we know that $a^2 + b^2 = 1$ and $ac + bd = 0$. Thus the two vectors (a, b) and (c, d) are orthogonal and $(a, b) = (\cos(t), \sin(t))$ for some $t \in \mathbb{R}$. This ultimately tells us that an orthogonal matrix is a rotation or a reflection – it represents a linear isometry. Therefore φ is surjective. Since φ is also injective (why?) it follows that it is a group isomorphism.

Example 2.4.8   The exponential function is a group isomorphism from $(\mathbb{R}, +)$ to $(\mathbb{R}_{>0}, \cdot)$. So $(\mathbb{R}, +)$ and $(\mathbb{R}_{>0}, \cdot)$ are isomorphic groups. This would have been impossible to prove without knowledge of the exponential function.

Notice that the kernel of the group homomorphism $G \to G/N$ is N and that the kernel of the determinant homomorphism from $GL_2(\mathbb{R})$ to $(\mathbb{R} \setminus \{0\}, \cdot)$ consists of matrices in $GL_2(\mathbb{R})$ with determinant 1.

We have the following general result on images and kernels of group homomorphisms.

Proposition 2.4.9   *Let $f : G \to K$ be a group homomorphism.*

(i) *The image $f(G) \subseteq K$ is a subgroup of K.*
(ii) *The kernel $\operatorname{Ker} f \subseteq G$ is a normal subgroup of G.*
(iii) *f is injective if and only if $\operatorname{Ker}(f) = \{e\}$.*

Proof.   First we prove that $f(G) = \{f(g) \mid g \in G\}$ is a subgroup of K. Since $f(e) = f(ee) = f(e)f(e)$, it follows that $e = f(e)$ by subsection 2.1.6. This shows that $e \in f(G)$. Let $x \in G$. Then $e = f(e) = f(xx^{-1}) = f(x)f(x^{-1})$ and $e = f(e) = f(x^{-1}x) = f(x^{-1})f(x)$. This shows that $f(x^{-1}) = f(x)^{-1}$. Thus if $f(x) \in f(G)$ then $f(x)^{-1} \in f(G)$. Finally if $f(x), f(y) \in f(G)$ then $f(x)f(y) = f(xy) \in f(G)$. This finishes the proof of (i).

Let us now prove that $\operatorname{Ker}(f)$ is a normal subgroup. We have already seen that $e \in \operatorname{Ker}(f)$ since $f(e) = e$. If $x \in \operatorname{Ker}(f)$ then $e = f(x) = f(x)^{-1} = f(x^{-1})$, showing that $x^{-1} \in \operatorname{Ker}(f)$. If $x, y \in \operatorname{Ker}(f)$ then $f(xy) = f(x)f(y) = ee = e$, showing that $xy \in \operatorname{Ker}(f)$. So $\operatorname{Ker}(f)$ is a subgroup of G. Let $N = \operatorname{Ker}(f)$. For every $g \in G$ and $x \in N$ we have $f((gx)g^{-1}) = (f(g)f(x))f(g^{-1}) = f(g)f(g)^{-1} = e$. This shows that $gNg^{-1} \subseteq N$. The inclusion $N \subseteq gNg^{-1}$ for every $g \in G$ follows from the fact that we have the inclusion $g^{-1}Ng \subseteq N$ for every $g \in G$. This finishes the proof of (ii).

Finally let us prove (iii). Since $f(e) = e$ it follows that $\mathrm{Ker}(f) = \{e\}$ if f is injective. Conversely, assume that $\mathrm{Ker}(f) = \{e\}$ and $f(x) = f(y)$. Then $f(y)^{-1}f(x) = f(y^{-1})f(x) = f(y^{-1}x) = e$. Therefore $y^{-1}x \in \mathrm{Ker}(f)$. This implies that $y^{-1}x = e$ or $x = y$. $\qquad\square$

2.5 The isomorphism theorem

Now suppose that N is a normal subgroup of G. How do we find out more about the quotient group G/N? The answer is that we identify the cosets G/N with some other known group using what is known as the isomorphism theorem.

Theorem 2.5.1 *Let G and K be groups and $f : G \to K$ a group homomorphism with kernel $N = \mathrm{Ker}(f)$. Then*

$$\tilde{f} : G/N \to f(G)$$

given by $\tilde{f}(gN) = f(g)$ is a well defined map and a group isomorphism.

Proof. First notice that $f(x) = f(y)$ if and only if $f(y)^{-1}f(x) = f(y^{-1}x) = e$ if and only if $y^{-1}x \in N$ for every $x, y \in G$. By Lemma 2.2.6(ii) this implies that $f(x) = f(y)$ if and only if $xN = yN$. We get thus that $\tilde{f}$ given by $\tilde{f}(gN) = f(g)$ is a well defined and injective map. It is a group homomorphism since

$$\begin{aligned}
\tilde{f}((g_1 N)(g_2 N)) &= \tilde{f}((g_1 g_2)N) \\
&= f(g_1 g_2) = f(g_1)f(g_2) \\
&= \tilde{f}(g_1 N)\tilde{f}(g_2 N)
\end{aligned}$$

for $g_1 N, g_2 N \in G/N$. It is surjective because f is surjective onto $f(G)$. Thus $\tilde{f}$ is a group isomorphism $G/N \to f(G)$. $\qquad\square$

One usually understands a quotient group G/N by finding a surjective group homomorphism $f : G \to K$ for a suitable group K such that $N = \mathrm{Ker}(f)$. Then Theorem 2.5.1 gives an isomorphism

$$\tilde{f} : G/N \xrightarrow{\sim} K.$$

Here are two examples of this.

Example 2.5.2 The subgroup $N = 2\pi\mathbb{Z} = \{2\pi m \mid m \in \mathbb{Z}\}$ of $(\mathbb{R}, +)$ is normal since $(\mathbb{R}, +)$ is an abelian group. What is $\mathbb{R}/N$? The strategy is to find a surjective group homomorphism $f : \mathbb{R} \to K$, with kernel $2\pi\mathbb{Z}$, onto some known group K. Here we can put $K = \{z \in \mathbb{C} \mid |z| = 1\}$, which is a group with multiplication as composition and use $f(x) = e^{ix}$ as the group homomorphism (recall that $e^{i(x+y)} = e^{ix}e^{iy}$). Then Ker $f = \{x \in \mathbb{R} \mid e^{ix} = 1\}$. Since $e^{ix} = \cos x + i \sin x$, this means that $x = 2\pi m$ for some $m \in \mathbb{Z}$. Now we can identify the quotient group $\mathbb{R}/2\pi\mathbb{Z}$ with the group of unit vectors in the complex plane by using the isomorphism

$$\tilde{f} : \mathbb{R}/2\pi\mathbb{Z} \xrightarrow{\sim} K$$

given in Theorem 2.5.1.

Example 2.5.3 Denote by A_3 the (normal) subgroup $\{e, d, f\}$ of S_3 (in the notation of Example 2.1.6). Then

$$S_3/A_3 \cong \mathbb{Z}/2\mathbb{Z},$$

since $|S_3/A_3| = |S_3|/|A_3| = 2$ and $\mathbb{Z}/2\mathbb{Z}$ is the only group of order 2 up to isomorphism. Can you construct an explicit surjective group homomorphism sgn : $S_3 \to \mathbb{Z}/2\mathbb{Z}$ such that Ker $(f) = A_3$?

2.6 Order of a group element

In a group G we can compose an element $g \in G$ with itself an arbitrary number of times $g, gg, (gg)g, \ldots$ Let us introduce the precise notion of powers of elements in groups. Define $g^0 = e$, $g^n = g^{n-1}g$ for $n > 0$ and $g^n = (g^{-1})^{-n}$ for $n < 0$ and every $g \in G$. Then we have a well defined map $f_g : \mathbb{Z} \to G$ given by $f_g(n) = g^n$.

Proposition 2.6.1 *Let G be a group and $g \in G$. The map*

$$f_g : \mathbb{Z} \to G$$

given by $f_g(n) = g^n$ is a group homomorphism from $(\mathbb{Z}, +)$ to G.

Proof. By the definition of g^n, where $n \in \mathbb{Z}$, we have $f_{g^{-1}}(-m) = f_g(m)$ for every $g \in G$, $m \in \mathbb{Z}$, along with $f_g(m+1) = f_g(m)f_g(1)$ and $f_g(m-1) = f_g(m)f_g(-1)$ for every $g \in G$, $m \geq 0$. This gives the identity $f_g(m+1) = f_g(m)f_g(1)$ for every $g \in G$, $m \in \mathbb{Z}$. From this we deduce that $f_g(m+n) = f_g(m)f_g(n)$ for every $g \in G$, $m \in \mathbb{Z}$ and $n \geq 0$. If $m < 0$ and $n < 0$ then $f_g(m+n) = f_{g^{-1}}(-m+(-n)) = f_{g^{-1}}(-m)f_{g^{-1}}(-n) = f_g(m)f_g(n)$.

This completes the proof that $f_g(m + n) = f_g(m) f_g(n)$ for every $m, n \in \mathbb{Z}$, showing that f_g is a group homomorphism. □

The image $f_g(\mathbb{Z}) = \{g^n \mid n \in \mathbb{Z}\}$ is denoted $\langle g \rangle$. It is an abelian subgroup of G (see Exercise 2.26). The number of elements in $\langle g \rangle$ is called the *order* of g. It is denoted ord (g). One usually thinks of the order of an element g as the smallest positive power of g giving the neutral element. If no such power exists, g is said to have infinite order.

Example 2.6.2   In the notation of Example 2.1.6, a has order 2 and f has order 3 in S_3. This follows from the composition table: $a \neq e$ but $a^2 = e$. Similarly $f \neq e$ and $f^2 = d \neq e$ but $f^3 = f^2 f = df = e$.

The element $[2] \in \mathbb{Z}/8\mathbb{Z}$ has order 4 and $[2] \in \mathbb{Z}/5\mathbb{Z}$ has order 5. However, the matrix

$$\begin{pmatrix} 1 & 1 \\ 0 & 1 \end{pmatrix}$$

has infinite order in the group $GL_2(\mathbb{R})$.

The following fundamental result turns out to be very useful for later computations in group theory (with applications to prime numbers and polynomials).

Proposition 2.6.3   *Let G be a finite group and let $g \in G$.*

(i) *The order* ord (g) *of g divides $|G|$.*
(ii) $g^{|G|} = e$.
(iii) *If $g^n = e$ for some $n > 0$ then* ord $(g) \mid n$.

Proof.   This is an application of Theorem 2.2.8. Let H denote the subgroup $\langle g \rangle$ generated by g. Since ord $(g) = |H|$, we get that $|G| = |G/H|$ ord (g). This proves (i). In the same way we have

$$g^{|G|} = g^{\text{ord}(g)|G/H|} = \left(g^{\text{ord}(g)}\right)^{|G/H|} = e^{|G/H|} = e.$$

This proves (ii). If $g^n = e$ then $n \in \text{Ker}(f_g)$. But $\text{Ker}(f_g) = n_g \mathbb{Z}$ and since $n > 0$ it follows that $n_g > 0$ and that g has finite order ord $(g) = n_g$. Since $n \in n_g \mathbb{Z} = \text{ord}(g)\mathbb{Z}$ we get that ord $(g) \mid n$. This proves (iii). □

2.7 Cyclic groups

Definition 2.7.1   *A cyclic group* is a group G containing an element g such that $G = \langle g \rangle$. The element g is called a *generator* of G (we say that G is *generated by g*).

Cyclic groups are very concrete objects. We can easily identify them with groups we know very well. Let $G = \langle g \rangle$ be a cyclic group and consider the group homomorphism $f_g : \mathbb{Z} \to G$. The kernel $\mathrm{Ker}(f_g)$ is a subgroup of $\mathbb{Z}$. Thus $\mathrm{Ker}(f_g) = n_g\mathbb{Z}$ for some unique natural number $n_g \geq 0$ by Proposition 2.2.3. By Theorem 2.5.1 we have a group isomorphism

$$\mathbb{Z}/n_g\mathbb{Z} \xrightarrow{\sim} \langle g \rangle = G.$$

This shows that a cyclic group is isomorphic to $\mathbb{Z}/n\mathbb{Z}$ for some $n \in \mathbb{N}$. Now we are in a position to illuminate the explicit computation in Example 2.1.9.

Proposition 2.7.2   *A group G of prime order $|G| = p$ is isomorphic to the cyclic group $\mathbb{Z}/p\mathbb{Z}$.*

Proof. Let $g \in G$ be an element in G different from the neutral element e. Then $f_g(\mathbb{Z})$ is a subgroup H of G with more than one element. Since $|H|$ divides $|G| = p$ (by Theorem 2.2.8) it follows that $|H| = |G|$ and therefore that $H = G$. This means that $f_g : \mathbb{Z} \to G$ is a surjective homomorphism. The kernel of f_g is $p\mathbb{Z}$ by Proposition 2.6.3. Now the result follows from Theorem 2.5.1. $\square$

Cyclic groups are in some sense the easiest groups to work with. Proposition 2.7.4 below tells almost the whole story about them. Before this let us go through an illustrative example.

Example 2.7.3   Let $[a] = a + 12\mathbb{Z}$, where $a \in \mathbb{Z}$. Then

$$\mathbb{Z}/12\mathbb{Z} = \{[0], [1], [2], [3], [4], [5], [6], [7], [8], [9], [10], [11]\}.$$

The order of $[3]$ is 4, since $\langle [3] \rangle = \{[0], [3], [6], [9]\}$. The orders of the elements in the group (appearing as above) are

$$1, 12, 6, 4, 3, 12, 2, 12, 3, 4, 6, 12$$

respectively. Notice that for every (natural) divisor d of 12, there is a unique subgroup of order d. This is the subgroup generated by $[12/d]$. Notice also that there are $\varphi(d)$ elements of order d.

Proposition 2.7.4 *Let G be a cyclic group.*

(i) *Every subgroup of G is cyclic.*
(ii) *Suppose that G is finite and that d is a divisor in $|G|$. Then G contains a unique subgroup H of order d.*
(iii) *There are $\varphi(d)$ elements of order d in G. These are the generators of H.*

Proof. If G is infinite then $G \cong \mathbb{Z}$. We know that every subgroup of $\mathbb{Z}$ has the form $d\mathbb{Z}$ for some $d \in \mathbb{N}$. Such a subgroup is cyclic and generated by d. Suppose that G is finite and that $|G| = N > 0$. We may assume that $G = \mathbb{Z}/N\mathbb{Z} = \{[0], [1], \ldots, [N-1]\}$. Let H be a subgroup of G. If $H \neq \{[0]\}$ we pick the smallest natural number $d > 0$ such that $[d] \in H$. If $[n] \in H$ then division with remainder gives $n = qd + r$, where $0 \leq r < d$. If $r > 0$ then $[n - qd] = [r] \in H$, contradicting the minimality of d. So $r = 0$ and $H = \langle[d]\rangle$. This proves (i).

Next, assume that d is a divisor in N. Let $m = N/d$. Then $[m]$ is an element of order d in G. If $[n]$ is another element of order d then $[dn] = [0]$. Thus $N \mid nd$ and so $m \mid n$. So every element in G of order d is some multiple of $[m]$. Since subgroups are cyclic, it follows that the only subgroup of order d is $H = \langle[m]\rangle$. This proves (ii).

Since H is the unique subgroup of order d, the elements of order d in G must be in one-to-one correspondence with the generators of H. We write $H = \{[0], [1], \ldots, [d-1]\}$ since $H \cong \mathbb{Z}/d\mathbb{Z}$. If $[a]$ is a generator of H then $\gcd(a, d) = 1$, because if $\gcd(a, d) = s > 1$, $a = bs$, $d = cs$ then we get $ca = cbs = bd$. Thus $[ca] = [0]$, where $1 \leq c < d$, contradicting that $[a]$ is a generator of H. However, if $\gcd(a, d) = 1$, $[a]$ has to be a generator of H: if $[ia] = 0$ then $d \mid ia$ and therefore $d \mid i$, since $\gcd(a, d) = 1$. This proves (iii).   □

Remark 2.7.5 In the notation of the proof of Proposition 2.7.4, the $\varphi(d)$ elements of order d in $\mathbb{Z}/N\mathbb{Z}$ are $\{[km] \mid 0 \leq k < N, \gcd(k, N) = 1\}$, where $m = N/d$.

Using the language of group theory we can now produce a very simple proof of an identity that seems related only to numbers.

Corollary 2.7.6   *Let N be a positive integer. Then*

$$\sum_{d \mid N} \varphi(d) = N,$$

where the sum is over $d \in \operatorname{div}(N)$.

Proof.   Let G be the cyclic group $\mathbb{Z}/N\mathbb{Z}$. Then

$$N = \sum_{g \in G} 1 = \sum_{d \mid N} \sum_{g \in G, \operatorname{ord}(g)=d} 1 = \sum_{d \mid N} \varphi(d)$$

by Proposition 2.7.4. $\square$

2.8 Groups and numbers

Let us see how Euler's theorem (Theorem 1.7.2) and the Chinese remainder theorem (Theorem 1.6.4) fit into the framework of groups.

2.8.1 Euler's theorem

Recall Euler's theorem. If a, n are relatively prime integers, where $n > 0$, then $a^{\varphi(n)} \equiv 1 \pmod{n}$. In the framework of groups we consider the finite group $G = (\mathbb{Z}/n\mathbb{Z})^*$ from subsection 2.3.2. The order of G is $\varphi(n)$. The integer a is relatively prime to n. Therefore $[a] \in G$. Now we can apply Proposition 2.6.3(ii) to obtain

$$[a]^{|G|} = [a]^{\varphi(n)} = [1].$$

This means that $a^{\varphi(n)} - 1 \in n\mathbb{Z}$ and therefore that $a^{\varphi(n)} \equiv 1 \pmod{n}$. You should really compare this with our original proof of Theorem 1.7.2. Where did all the computations go? The answer is that groups form another level of abstraction. Proofs become simpler.

Before moving on to the group version of the Chinese remainder theorem we need to define product groups.

2.8.2 Product groups

If $G_1, G_2, \ldots, G_n$ are groups then the product

$$G = G_1 \times G_2 \times \cdots \times G_n = \{(g_1, g_2, \ldots, g_n) \mid g_1 \in G_1, g_2 \in G_2, \ldots, g_n \in G_n\}$$

has the natural composition

$$(g_1, g_2, \ldots, g_n)(h_1, h_2, \ldots, h_n) = (g_1 h_1, g_2 h_2, \ldots, g_n h_n).$$

You can easily check that this composition is associative (it is associative at each component). The neutral element is $(e, e, \ldots, e)$ and the inverse of the group element $g = (g_1, \ldots, g_n)$ is $g^{-1} = (g_1^{-1}, \ldots, g_n^{-1})$. So G is a group called the *product group* of $G_1, \ldots, G_n$. Also, if H is a group and we have group homomorphisms $\varphi_i : H \to G_i, i = 1, \ldots, n$, then

$$\varphi(g) = (\varphi_1(g), \ldots, \varphi_n(g))$$

is a group homomorphism from H to $G = G_1 \times \cdots \times G_n$. Before giving the group version of the Chinese remainder theorem, let us record the following lemma on product groups.

Lemma 2.8.1 *Let M, N be normal subgroups of a group G with $M \cap N = \{e\}$. Then MN is a subgroup of G and*

$$\pi : M \times N \to MN$$

given by $\pi(x, y) = xy$ is an isomorphism.

Proof. Lemma 2.3.6 tells us that MN is a subgroup. In order for π to be a homomorphism we must prove that $(xy)(x'y') = (xx')(yy')$, where $x, x' \in M$ and $y, y' \in N$. This is seen by rewriting $(xy)(x'y')$ as $(xx')(x'^{-1}yx'y^{-1})(yy')$ and noticing that $x'^{-1}yx'y^{-1} \in M \cap N = \{e\}$, since M and N are normal subgroups of G. Since the kernel of π is isomorphic to $M \cap N$ and the image of π is MN, π has to be bijective and therefore an isomorphism. $\qquad \square$

2.8.3 The Chinese remainder theorem

Here is the group version of the Chinese remainder theorem (Theorem 1.6.4).

Proposition 2.8.2 *Let $n_1, \ldots, n_r \in \mathbb{Z}$ be pairwise relative prime integers and let $N = n_1 \cdots n_r$. If φ_i denotes the canonical group homomorphism $\pi_{n_i \mathbb{Z}} : \mathbb{Z} \to \mathbb{Z}/n_i \mathbb{Z}$ then the map*

$$\tilde{\varphi} : \mathbb{Z}/N\mathbb{Z} \to \mathbb{Z}/n_1\mathbb{Z} \times \cdots \times \mathbb{Z}/n_r\mathbb{Z}$$

given by $\varphi(x + N\mathbb{Z}) = (\varphi_1(x), \ldots, \varphi_r(x))$ is a group isomorphism.

Proof. The map

$$\varphi : \mathbb{Z} \to \mathbb{Z}/n_1\mathbb{Z} \times \cdots \times \mathbb{Z}/n_r\mathbb{Z}$$

given by $\varphi(x) = (\varphi_1(x), \ldots, \varphi_r(x))$ is a group homomorphism by subsection 2.8.2. If $n \in \text{Ker}(\varphi)$ then $\varphi_1(n) = 0, \ldots, \varphi_r(n) = 0$. This means that $n \in n_1\mathbb{Z}, \ldots, n \in n_r\mathbb{Z}$ or that $n_1 \mid n, \ldots, n_r \mid n$. By Corollary 1.5.11 we get that $N = n_1 \cdots n_r \mid n$ so that $n \in N\mathbb{Z}$. This proves that $\text{Ker}(\varphi) \subseteq N\mathbb{Z}$. The other inclusion is left to the reader. Now Theorem 2.5.1 tells us that we have an isomorphism

$$\tilde{\varphi} : \mathbb{Z}/N\mathbb{Z} \to \varphi(\mathbb{Z}) \subseteq \mathbb{Z}/n_1\mathbb{Z} \times \cdots \times \mathbb{Z}/n_r\mathbb{Z}.$$

But since the number of elements in $\mathbb{Z}/N\mathbb{Z}$ equals the number of elements in $\mathbb{Z}/n_1\mathbb{Z} \times \cdots \times \mathbb{Z}/n_r\mathbb{Z}$, we get that $\varphi(\mathbb{Z}) = \mathbb{Z}/n_1\mathbb{Z} \times \cdots \times \mathbb{Z}/n_r\mathbb{Z}$ and $\tilde{\varphi}$ is thus an isomorphism. $\square$

Using the notation of Proposition 2.8.2 we have actually proved that

$$\mathbb{Z}/n_1\mathbb{Z} \times \cdots \times \mathbb{Z}/n_r\mathbb{Z}$$

is a cyclic group $\cong \mathbb{Z}/N\mathbb{Z}$.

Example 2.8.3 The product group $G = \mathbb{Z}/2\mathbb{Z} \times \mathbb{Z}/2\mathbb{Z}$ is not cyclic, since the maximal order of an element is 2. One may prove that $(\mathbb{Z}/8\mathbb{Z})^* \cong \mathbb{Z}/2\mathbb{Z} \times \mathbb{Z}/2\mathbb{Z}$.

2.9 Symmetric and alternating groups

In Example 2.1.6 we constructed the group S_3 of bijective maps of a set M of three elements to itself. The composition in S_3 is the composition of maps. The bijective map given by $1 \mapsto 2, 2 \mapsto 3$ and $3 \mapsto 1$ was denoted

$$\begin{pmatrix} 1 & 2 & 3 \\ 2 & 3 & 1 \end{pmatrix}.$$

Of course, the same construction makes sense for a set $M_n = \{1, 2, \ldots, n\}$ with n elements and this leads to the so-called *symmetric group* S_n on n elements. Thus S_n consists of the bijective maps from M_n to itself. It is a group with composition of maps as the composition, and one may show that $|S_n| = n!$ by counting permutations of the numbers $1, \ldots, n$. The elements (bijective maps) of S_n are called *permutations*. As in the S_3 setting, a bijective map $\sigma \in S_n$ will be denoted

$$\begin{pmatrix} 1 & 2 & \dots & n \\ \sigma(1) & \sigma(2) & \dots & \sigma(n) \end{pmatrix}.$$

Symmetric groups are in general non-abelian. We have for example

$$\begin{pmatrix} 1 & 2 & 3 \\ 2 & 1 & 3 \end{pmatrix} \circ \begin{pmatrix} 1 & 2 & 3 \\ 1 & 3 & 2 \end{pmatrix} \neq \begin{pmatrix} 1 & 2 & 3 \\ 1 & 3 & 2 \end{pmatrix} \circ \begin{pmatrix} 1 & 2 & 3 \\ 2 & 1 & 3 \end{pmatrix},$$

since the map on the left hand side assumes the value 3 at 2 and the map on the right hand side assumes the value 1 at 2. In an important special case one can actually prove that $\sigma \tau = \tau \sigma$, where σ, τ are certain permutations in S_n.

Definition 2.9.1   Suppose that $\sigma \in S_n$. Then we define

$$M_\sigma = \{x \in M_n \mid \sigma(x) \neq x\}.$$

Permutations $\sigma, \tau \in S_n$ are called disjoint if $M_\sigma \cap M_\tau = \emptyset$.

One may say loosely that disjoint permutations move different numbers. They have the following pleasant property.

Proposition 2.9.2   *Let $\sigma, \tau \in S_n$ be disjoint permutations in S_n. Then*

$$\sigma \tau = \tau \sigma.$$

Proof.   We must prove that $\sigma(\tau(x)) = \tau(\sigma(x))$ for every $x \in M_n$. If $x \notin M_\sigma \cup M_\tau$ then $\sigma(x) = x$ and $\tau(x) = x$ and both sides are equal to x. If $x \in M_\sigma$ then $\sigma(x) \in M_\sigma$ (why?). Therefore we have $\tau(\sigma(x)) = \sigma(x)$ and similarly $\sigma(\tau(x)) = \sigma(x)$. So both sides are equal in this case. The case $x \in M_\tau$ is treated in the same way.   □

2.9.1 Cycles

Some permutations in S_n deserve special attention. Suppose we are given k different elements $x_1, x_2, \dots, x_k$ of M_n. A permutation $\sigma \in S_n$ given by

$$\sigma(x_1) = x_2, \quad \sigma(x_2) = x_3, \quad \dots, \quad \sigma(x_{k-1}) = x_k, \quad \sigma(x_k) = x_1$$

and $\sigma(x) = x$ if $x \notin \{x_1, \dots, x_k\}$ is called a *k-cycle*. It is denoted

$$\sigma = (x_1 \, x_2 \dots x_k)$$

to indicate that $x_2 = \sigma(x_1), \ldots, x_1 = \sigma(x_k)$. In this notation σ may be written in the following k different ways:

$$(x_1 \, x_2 \, \ldots \, x_{k-1} \, x_k),$$
$$(x_2 \, x_3 \, \ldots \, x_k \, x_1),$$

$$\vdots$$

$$(x_k \, x_1 \, \ldots \, x_{k-2} \, x_{k-1}).$$

Notice that $M_\sigma = \{x_1, x_2, \ldots, x_k\}$ and that the order of a k-cycle in S_n is k.

Example 2.9.3   Consider

$$\begin{pmatrix} 1 & 2 & 3 \\ 2 & 3 & 1 \end{pmatrix}.$$

This is the 3-cycle $(1\,2\,3)$ in S_3. As an element in the group S_3 it has order 3. Notice that $(1\,2\,3) = (2\,3\,1) = (3\,1\,2)$.

A 1-cycle is literally translated as the identity map. A 2-cycle is called a *transposition*. Notice that a transposition is its own inverse in S_n. A transposition of the form $s_i = (i \ \ i+1)$, where $i = 1, \ldots, n-1$, is called a *simple transposition*.

Example 2.9.4   In S_3 we have

$$(1\,2\,3) = \begin{pmatrix} 1 & 2 & 3 \\ 2 & 3 & 1 \end{pmatrix} = s_1 s_2.$$

This follows by evaluating $(1\,2\,3)$ and the composition $s_1 s_2$ on $1, 2$ and 3 and seeing that they give the same result.

It turns out that every permutation can be expressed as a product of disjoint cycles. Such an expression is useful, for example, in the following proposition.

Proposition 2.9.5   *Let* $\sigma \in S_n$ *be written as a product of disjoint cycles* $\sigma_1 \cdots \sigma_r$. *Then the order of* σ *is the least common multiple of the orders of the cycles* $\sigma_1, \ldots, \sigma_r$.

Proof.   Since $\sigma_i \sigma_j = \sigma_j \sigma_i$ when $i \neq j$ we get $\sigma^n = \sigma_1^n \cdots \sigma_r^n$ for $n \in \mathbb{N}$. If $\sigma^n = e$ then $\sigma_i^n = e$ for $i = 1, \ldots, r$, as $\sigma_1^n, \ldots, \sigma_r^n$ are disjoint permutations. Therefore n is divisible by the orders of the cycles, by Proposition

2.6.3(iii). This means that the least common multiple m of the orders of the cycles is $\leq \mathrm{ord}\,(\sigma)$. However, $\sigma_i^m = e$ for every $i = 1, \ldots, r$. Therefore $\mathrm{ord}\,(\sigma) = m$. □

We have the following fundamental proposition.

Proposition 2.9.6 *Every permutation $\sigma \in S_n$ is a product of unique disjoint cycles.*

Proof. The proof of the existence uses induction on the number of elements in M_σ. If $|M_\sigma| = 0$ then σ is a product of disjoint 1-cycles. Assume that $|M_\sigma| > 0$. Pick $x \in M_\sigma$. Then $x \neq \sigma(x)$. Form the sequence $x = \sigma^0(x), \sigma(x), \sigma^2(x), \ldots$ of elements in M_n and stop when you encounter the first repetition $\sigma^k(x)$, where $\sigma^k(x) = \sigma^j(x)$ for some $0 \leq j < k$. Then $j = 0$ (why?). Define the cycle $\tau = (x_1 \, x_2 \, \ldots \, x_k)$ by

$$x_1 = x, \quad x_2 = \sigma(x_1), \quad \ldots, \quad x_k = \sigma(x_{k-1}) \quad \text{and} \quad x_1 = \sigma(x_k).$$

Now $M_{\sigma\tau^{-1}} = M_\sigma \setminus \{x_1, \ldots, x_k\}$, because if $x \notin \{x_1, \ldots, x_k\}$ then $\tau^{-1}(x) = x$. Such an x will satisfy $\sigma(\tau^{-1}(x)) \neq x$ if and only if $\sigma(x) \neq x$. However, if $x \in \{x_1, \ldots, x_k\}$ then $\sigma(x) \neq x$ but $\sigma(\tau^{-1}(x)) = x$, since x can be written $\tau(y)$ for $y \in \{x_1, \ldots, x_k\}$ with $\sigma(y) = x$. By induction $\sigma\tau^{-1}$ is a product of disjoint cycles $\tau_1 \cdots \tau_r$. Since τ must be disjoint from $\tau_1, \ldots, \tau_r$, it follows that

$$\sigma = \tau_1 \cdots \tau_r \tau$$

is a product of disjoint cycles. This proves that a permutation can be written as a product of disjoint cycles. The uniqueness part can be deduced from the existence proof. In fact, if $\sigma = \sigma_1 \cdots \sigma_r$ is written as a product of disjoint cycles $\sigma_1, \ldots, \sigma_r$ then $M_\sigma = M_{\sigma_1} \cup \cdots \cup M_{\sigma_r}$ and $M_{\sigma_i} \cap M_{\sigma_j} = \emptyset$ if $i \neq j$, since σ_i and σ_j are disjoint permutations if $i \neq j$. So, if $x \in M_\sigma$ then $x \in M_{\sigma_j}$ for a unique $j = 1, \ldots, r$ and $\sigma_j = (x \, \sigma(x) \, \sigma^2(x) \, \ldots)$ since $\sigma(x) = \sigma_j(x)$, when $x \in M_{\sigma_j}$. In this way the cycles occurring in σ written as a product of disjoint cycles are uniquely determined by σ. □

Example 2.9.7 The element

$$\sigma = \begin{pmatrix} 1 & 2 & 3 & 4 & 5 & 6 \\ 6 & 3 & 1 & 5 & 4 & 2 \end{pmatrix} \in S_6$$

can be written as the product $(1623)(45)$ of disjoint cycles. One simply mimics the procedure outlined in the proof of Proposition 2.9.6: $\sigma(1) = 6$, $\sigma(6) = 2$,

$\sigma(2) = 3$, $\sigma(3) = 1$ gives the 4-cycle (1623) and $\sigma(4) = 5$, $\sigma(5) = 4$ gives the transposition (45). The order of σ is lcm(2, 4) = 4.

The following lemma will be very important for later computations.

Lemma 2.9.8 *Suppose that $\tau = (i_1 i_2 \dots i_k)$ is a k-cycle and σ a permutation in S_n. Then*

$$\sigma(i_1 i_2 \dots i_k)\sigma^{-1} = (\sigma(i_1)\sigma(i_2)\dots \sigma(i_k)).$$

Proof. Let $J = \{\sigma(i_1), \dots, \sigma(i_k)\}$. Then the left and right hand sides assume the same value on $i \in J$. Since they both map $i \notin J$ to itself, they must be the same permutations.   □

2.9.2 Simple transpositions and "bubble sort"

Let us describe one of the simplest sorting algorithms ("bubble sort") for sorting n numbers $a_1, \dots, a_n$. You run through the list $a_1, \dots, a_n$. Each time you encounter a neighboring pair $a_i > a_{i+1}$ that is not in (ascending) order, you switch the two numbers and go back to the beginning. Do this until there are no more unordered neighboring pairs. Then the sequence has been sorted into ascending order. How does this relate to permutations? Take a look at the example below.

Example 2.9.9 Suppose that we consider the permutation 631542 of the sequence 123456. Using "bubble sort" you can reorder the permuted sequence by switching neighbors:

$$\begin{array}{lllll}
631542 & 361542 & 316542 & 136542 & 135642 \\
135462 & 134562 & 134526 & 134256 & 132456 \\
123456. & & & &
\end{array}$$

The process of switching neighbors corresponds to the simple transpositions

$$(12)(23)(12)(34)(45)(34)(56)(45)(34)(23),$$

where the numbers refer to the positions in the sequence. In the language of permutations and S_6 you may express the first step of the bubble sort as

$$\begin{pmatrix} 1 & 2 & 3 & 4 & 5 & 6 \\ 6 & 3 & 1 & 5 & 4 & 2 \end{pmatrix}(12) = \begin{pmatrix} 1 & 2 & 3 & 4 & 5 & 6 \\ 3 & 6 & 1 & 5 & 4 & 2 \end{pmatrix}.$$

In total we have proved that

$$\begin{pmatrix} 1 & 2 & 3 & 4 & 5 & 6 \\ 6 & 3 & 1 & 5 & 4 & 2 \end{pmatrix}(12)(23)(12)(34)(45)(34)(56)(45)(34)(23)$$

$$= \begin{pmatrix} 1 & 2 & 3 & 4 & 5 & 6 \\ 1 & 2 & 3 & 4 & 5 & 6 \end{pmatrix}$$

and therefore that

$$\begin{pmatrix} 1 & 2 & 3 & 4 & 5 & 6 \\ 6 & 3 & 1 & 5 & 4 & 2 \end{pmatrix} = (23)(34)(45)(56)(34)(45)(34)(12)(23)(12).$$

You should check this by evaluating the permutations on the left and right hand side on 1, 2, 3, 4, 5 and 6.

Example 2.9.9 illustrates the result that every permutation is a product of simple transpositions. What is the minimal number of simple transpositions needed for writing a permutation σ as a product in this way? Surprisingly, the answer lies in counting the number of ordered pairs (i, j), $i < j$, for which the values $\sigma(i) > \sigma(j)$ are in the wrong order. This is the reasoning behind the following definition.

Definition 2.9.10 Let $\sigma \in S_n$ be a permutation. A pair of indices (i, j), where $1 \le i < j \le n$, is called an *inversion* (of σ) if $\sigma(i) > \sigma(j)$. Let

$$I_\sigma = \{(i, j) \mid 1 \le i < j \le n \text{ and } \sigma(i) > \sigma(j)\}$$

denote the set of inversions and let $n(\sigma) = |I_\sigma|$ be the number of inversions of σ.

Example 2.9.11 We have that

$$n\left(\begin{pmatrix} 1 & 2 & 3 \\ 2 & 3 & 1 \end{pmatrix}\right) = 2,$$

since $(1, 3)$ and $(2, 3)$ are the only inversions (corresponding to $2 > 1$ and $3 > 1$). Again counting inversions we find that

$$n\left(\begin{pmatrix} 1 & 2 & 3 & 4 & 5 & 6 \\ 6 & 3 & 1 & 5 & 4 & 2 \end{pmatrix}\right) = 10.$$

This agrees with the number of simple transpositions we found in Example 2.9.9.

Proposition 2.9.12   *The permutation $\sigma \in S_n$ is the identity map if and only if $n(\sigma) = 0$. If σ is not the identity map then there exists $i = 1, \ldots, n - 1$ such that $\sigma(i) > \sigma(i + 1)$.*

Proof.   If σ is the identity map then it has no inversions. Therefore $n(\sigma) = 0$. If $n(\sigma) = 0$ and σ is not the identity map then there exists a smallest $i \in M_n$ such that $\sigma(i) > i$. The pair $(i, \sigma^{-1}(i))$ is an inversion for σ, contradicting that $n(\sigma) = 0$. If σ is a permutation satisfying $\sigma(1) < \sigma(2) < \cdots < \sigma(n)$ then σ has to be the identity map, since $n(\sigma) = 0$. This proves the last part of the proposition. $\square$

The following lemma is crucial.

Lemma 2.9.13   *Let $s_i \in S_n$ be a simple transposition and $\sigma \in S_n$. Then*

$$n(\sigma s_i) = \begin{cases} n(\sigma) + 1 & \text{if } \sigma(i) < \sigma(i + 1), \\ n(\sigma) - 1 & \text{if } \sigma(i) > \sigma(i + 1). \end{cases}$$

Proof.   Assume that $\sigma(i) < \sigma(i + 1)$. Since $(i, i + 1)$ is an inversion for σs_i (why?) we only need to establish a bijective map

$$\varphi : I_\sigma \to I_{\sigma s_i} \setminus \{(i, i + 1)\}.$$

Such a bijective map is given by $\varphi((k, l)) = (s_i(k), s_i(l))$. If $(k, l) \in I_\sigma$ then $s_i(k) < s_i(l)$, because the only way this can fail is if $k = i$ and $l = i + 1$ and, by assumption, $(i, i + 1) \notin I_\sigma$. Now $(s_i(k), s_i(l)) \in I_{\sigma s_i}$, since $(k, l) \in I_\sigma$. In the same way, if $(k, l) \in I_{\sigma s_i} \setminus \{(i, i + 1)\}$ then $(s_i(k), s_i(l)) \in I_\sigma$. This proves that φ is a bijective map. If $\sigma(i) > \sigma(i + 1)$ then we work with the permutation σs_i. In this case we know that $(\sigma s_i)(i) < (\sigma s_i)(i + 1)$ and therefore it follows that $n((\sigma s_i)s_i) = n(\sigma) = n(\sigma s_i) + 1$ by what we have already proved. $\square$

Proposition 2.9.14   *Let $\sigma \in S_n$. Then*

(i) *σ is a product of $n(\sigma)$ simple transpositions.*
(ii) *$n(\sigma)$ is the minimal number of simple transpositions needed in writing σ as a product of simple transpositions.*

Proof.   We will use induction on $n(\sigma)$ for proving (i). If $n(\sigma) = 0$ then σ is the identity map by Proposition 2.9.12 and we are done (σ is the empty product of simple transpositions, which is the identity by definition). If not, we may find $i = 1, \ldots, n - 1$ such that $\sigma(i) > \sigma(i + 1)$ according to Proposition 2.9.12. Then $n(\sigma s_i) = n(\sigma) - 1$ by Lemma 2.9.13. By induction $\eta = \sigma s_i$ can be written as a product of $n(\sigma) - 1$ simple transpositions. Then $\sigma = \eta s_i$ is a product of $n(\sigma)$ simple transpositions. This proves (i).

Let $\ell(\sigma)$ denote the minimal number of simple transpositions needed in writing σ as a product of simple transpositions. Then $n(\sigma) \geq \ell(\sigma)$ by (i). We will prove that $\ell(\sigma) = n(\sigma)$ using induction on $\ell(\sigma)$. The case $\ell(\sigma) = 0$ follows as in the proof of (i). Assume that $\ell(\sigma) > 0$. Then we may find a simple transposition s_i such that $\ell(\sigma s_i) = \ell(\sigma) - 1$. Therefore $\ell(\sigma s_i) = n(\sigma s_i)$ by induction and $\ell(\sigma) \geq n(\sigma)$ by Lemma 2.9.13. This proves (ii). □

2.9.3 The alternating group

Definition 2.9.15 The sign of a permutation $\sigma \in S_n$ is

$$\text{sgn}(\sigma) = (-1)^{n(\sigma)}.$$

A permutation with positive sign is called even. A permutation with negative sign is called odd.

Proposition 2.9.16 *The sign*

$$\text{sgn} : S_n \to \{\pm 1\}$$

of a permutation is a group homomorphism, where the composition in $\{\pm 1\}$ is multiplication.

Proof. We must prove that $\text{sgn}(\sigma\tau) = \text{sgn}(\sigma)\text{sgn}(\tau)$, where $\sigma, \tau \in S_n$. Since τ is a product of simple transpositions we may assume that τ itself is a simple transposition s_i. By Lemma 2.9.13 we have $n(\sigma s_i) = n(\sigma) \pm 1$, so that $\text{sgn}(\sigma s_i) = -\text{sgn}(\sigma)$. Thus $\text{sgn}(\sigma s_i) = \text{sgn}(\sigma)\text{sgn}(s_i)$, as $n(s_i) = 1$. □

The set of even permutations in S_n is denoted A_n and called the *alternating group*. It follows by Proposition 2.9.16 that A_n is a normal subgroup of S_n, being the kernel of sgn. By Theorem 2.5.1, we get the group isomorphism

$$S_n/A_n \xrightarrow{\sim} \{\pm 1\},$$

showing that $|A_n| = |S_n|/2 = n!/2$ for $n > 1$. Before moving on, let us see how one can determine $\text{sgn}(\sigma)$ for a permutation $\sigma \in S_n$ from its disjoint cycle decomposition (Proposition 2.9.6). Since sgn is a group homomorphism, we only need to compute the sign of a cycle.

Proposition 2.9.17 *Let $n \geq 2$. A transposition $\tau = (i\ j) \in S_n$ is an odd permutation. The sign of an r-cycle $\sigma = (x_1\ x_2\ \ldots\ x_r) \in S_n$ is $(-1)^{r-1}$.*

Proof. We can find a permutation $\eta \in S_n$ such that $\eta(1) = i$ and $\eta(2) = j$. This implies that $-1 = \mathrm{sgn}(1\ 2) = \mathrm{sgn}(\eta(1\ 2)\eta^{-1}) = \mathrm{sgn}((\eta(1)\ \eta(2))) = \mathrm{sgn}(\tau)$. Therefore τ is odd. To see that the sign of σ is $(-1)^{r-1}$, we simply write

$$(x_1\ x_2\ \ldots\ x_r) = (x_1\ x_2)(x_2\ x_3)\ldots(x_{r-1}x_r).$$

We have expressed σ as a product of $r - 1$ transpositions. Therefore $\mathrm{sgn}(\sigma) = (-1)^{r-1}$. □

2.9.4 Simple groups

A group N is called *simple* if $\{e\}$ and N are the only normal subgroups of N. One can prove that any finite group G has a decreasing sequence of subgroups,

$$G = G_0 \supseteq G_1 \supseteq G_2 \supseteq \cdots \supseteq G_{n-1} \supseteq G_n = \{e\},$$

such that G_{i+1} is a normal subgroup of G_i and the quotient group G_i/G_{i+1} is a simple finite group. One may also prove that the simple quotient groups occurring in such a decreasing sequence are uniquely determined up to isomorphism. In this sense the simple finite groups form the building blocks for all finite groups.

Here we prove the following classical result due to E. Galois (1811–32). When developed a little further, into Galois theory, it accounts for the miraculous fact that there is no formula (involving the usual arithmetical operations and extracting roots) for the solution of a general algebraic equation of degree ≥ 5. First we need a simple but important lemma.

Lemma 2.9.18 *Every permutation in A_n is a product of 3-cycles if $n \geq 3$.*

Proof. A permutation in A_n is a product of an even number of transpositions. Consider four distinct numbers a, b, c and d. Then $(a\ b)(c\ d) = (a\ d\ c)(a\ b\ c)$ and $(a\ b)(b\ c) = (a\ b\ c)$. So we may replace consecutive pairs of transpositions with products of 3-cycles. This proves the claim. □

Theorem 2.9.19 *The alternating group A_n is simple for $n \geq 5$.*

Proof. We will prove that

(i) Given a 3-cycle $\tau \in A_n$, there is a permutation $\sigma \in A_n$ such that $\sigma \tau \sigma^{-1} =$ (123).

(ii) A non-trivial normal subgroup N of A_n must contain a 3-cycle.

We now go through these two steps. (i) Let $\tau = (ijk)$ be a 3-cycle. We can find a permutation $\sigma \in S_n$ such that $\sigma(i) = 1$, $\sigma(j) = 2$ and $\sigma(k) = 3$. Now Lemma 2.9.8 gives

$$\sigma(ijk)\sigma^{-1} = (123).$$

We may assume that $\sigma \in A_n$ by replacing σ with $(45)\sigma$ in the case $\sigma \notin A_n$: $((45)(123)(45)^{-1} = (123))$. This proves (i).

(ii) Let N be a non-trivial normal subgroup of A_n. We need to show that N contains a 3-cycle τ. Let $\sigma \in N$ denote an element $\neq e$. Write σ as a product $\tau_1 \tau_2 \cdots \tau_r$ of disjoint cycles. If two of the disjoint cycles are transpositions, we may assume that $\tau_1 = (12)$ and $\tau_2 = (34)$ and thus $\sigma = (12)(34)\eta$ for some $\eta \in S_n$. Putting $\tau = (123)$ we get a new permutation $\sigma_1 = \tau \sigma \tau^{-1} \sigma^{-1}$ that also lies in N, since N is a normal subgroup. Composing permutations we get (using Lemma 2.9.8)

$$\sigma_1 = \tau \sigma \tau^{-1} \sigma^{-1} = (13)(24).$$

Now using the same trick with $\rho = (245)$, we get

$$\sigma_2 = \rho \sigma_1 \rho^{-1} \sigma_1^{-1} = (254).$$

So σ_2 is the desired 3-cycle in N. If σ contains a cycle $(1234\ldots)$ of length at least 4 we get the 3-cycle

$$\tau \sigma \tau^{-1} \sigma^{-1} = (124)$$

in N, where $\tau = (123)$. The only case left is where σ contains a 3-cycle (123) and another cycle $(45\ldots)$. In this case we get, using $\tau = (234)$, that

$$\tau \sigma \tau^{-1} \sigma^{-1} = (14235),$$

which is a a cycle of length 5. We already know why this implies that N contains a 3-cycle.

We know that, for every 3-cycle τ, there is an element $\sigma \in A_n$ such that

$$\sigma \tau \sigma^{-1} = (123).$$

This means that, given two arbitrary 3-cycles τ_1 and τ_2, there is a $\sigma \in A_n$ such that $\sigma \tau_1 \sigma^{-1} = \tau_2$. Thus if a normal subgroup N of A_n contains just one 3-cycle, it will have to contain all 3-cycles! In this way we have proved that a normal subgroup N of A_n is either $\{e\}$ or A_n, as we know that every element of A_n is a product of 3-cycles by Lemma 2.9.18.   □

One of the milestones of modern group theory is the theorem of Feit and Thompson. They proved in 1963 that the order of a non-abelian finite simple group must be even. The proof of this takes up more than 250 pages [9]. Simple finite groups fall into some well defined families except for 26 finite simple groups, the sporadic groups. The largest sporadic group is called the monster group. It has

$$808017424794512875886459904961710757005754368000000000$$
$$= 2^{46} \cdot 3^{20} \cdot 5^9 \cdot 7^6 \cdot 11^2 \cdot 13^3 \cdot 17 \cdot 19 \cdot 23 \cdot 29 \cdot 31 \cdot 41 \cdot 47 \cdot 59 \cdot 71$$

elements. The classification of the finite simple groups was completed in 1980 (but has not yet been written up completely!).

2.9.5 The 15-puzzle

Can you interchange the empty square successively with adjacent squares so that the configuration on the left gets changed into the "correct" configuration, the one on the right below?

1	2	3	4
5	6	7	8
9	10	11	12
13	15	14	

1	2	3	4
5	6	7	8
9	10	11	12
13	14	15	

This is the classical 15-puzzle, as published by the American puzzlemaker Sam Loyd in 1878. He offered a prize of 1000 dollars for the first correct solution to the problem. He went on to write (see [10])

> People became infatuated with the puzzle and ludicrous tales are told of shopkeepers who neglected to open their stores; of a distinguished clergyman who stood under a street lamp all through a wintry night trying to recall the way he had performed the feat. The mysterious feature of the puzzle is that none seem to be able to remember the sequence of moves whereby they feel sure they succeeded in solving the puzzle. Pilots are said to have wrecked their ships, and engineers rush

their trains past stations. A famous Baltimore editor tells how he went for his noon lunch and was discovered by his frantic staff long past midnight pushing little pieces of pie around on a plate! Farmers are known to have deserted their plows ...

The frustrated farmer below appeared in Loyd's original article on the puzzle.

Following [2] we will go through a method of analyzing this problem using symmetric and alternating groups. Each square (including the empty square) occupies one of the 16 numbered *cells* below.

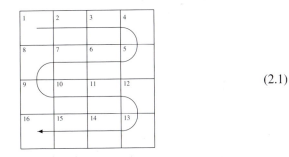

(2.1)

A configuration C maps to a permutation σ_C in S_{15} defined by writing the squares according to their order along the snake pattern in (2.1) (forgetting the empty square). For example, the "correct" configuration maps to

$$\begin{pmatrix} 1 & 2 & 3 & 4 & 5 & 6 & 7 & 8 & 9 & 10 & 11 & 12 & 13 & 14 & 15 \\ 1 & 2 & 3 & 4 & 8 & 7 & 6 & 5 & 9 & 10 & 11 & 12 & 15 & 14 & 13 \end{pmatrix}$$
$$= (5\ 8)(6\ 7)(13\ 15)$$

and the "evil" Loyd configuration maps to

$$\begin{pmatrix} 1 & 2 & 3 & 4 & 5 & 6 & 7 & 8 & 9 & 10 & 11 & 12 & 13 & 14 & 15 \\ 1 & 2 & 3 & 4 & 8 & 7 & 6 & 5 & 9 & 10 & 11 & 12 & 14 & 15 & 13 \end{pmatrix}$$
$$= (5\ 8)(6\ 7)(13\ 14\ 15).$$

The correct configuration maps to an odd permutation and the "evil" Loyd configuration maps to an even permutation. We will see shortly that this is the reason why the original 15-puzzle was unsolvable, so that Loyd was sure never to lose his 1000 dollars.

The mapping of a configuration to a permutation in S_{15} using the snake pattern is not a one-to-one correspondence. The configurations you get by moving the empty square along the snake all map to the same permutation in S_{15}. If C is a configuration where the blank square occupies cell b and square i occupies cell j then

$$\sigma_C^{-1}(i) = \begin{cases} j & \text{if } j < b, \\ j - 1 & \text{if } j > b. \end{cases}$$

Suppose that we have a configuration C_1 where the empty square occupies cell i. By moving the empty square (legally) to cell j we get a new configuration C_2, where the blank square occupies cell j. Then σ_{C_1} and σ_{C_2} are related through a fixed permutation $\sigma_{i,j} \in S_{15}$ via $\sigma_{C_2}^{-1} = \sigma_{i,j}\sigma_{C_1}^{-1}$. The permutations corresponding to the legal moves can be read off from (2.1). They are

$$\sigma_{1,2}, \sigma_{2,3}, \ldots, \sigma_{15,16}, \sigma_{1,8}, \sigma_{2,7}, \sigma_{3,6}, \sigma_{7,10}, \sigma_{6,11}, \sigma_{5,12}, \sigma_{11,14}, \sigma_{10,15}, \sigma_{9,16}$$

$$(2.2)$$

along with their inverse permutations. It is easy to see that $\sigma_{1,2} = \sigma_{2,3} = \cdots = \sigma_{15,16} = 1$. These moves do not affect σ_C for a given configuration C for which the empty square is positioned in the appropriate cell. Let us have a closer look at $\sigma_{1,8}$. After having done this move, which consists in moving the empty square from cell 1 to cell 8, the square that was number 1 becomes number 2, the square that was number 2 becomes number 3, ... , the square that was number 7 becomes number 1. This proves that

$$\sigma_{1,8} = \begin{pmatrix} 1 & 2 & 3 & 4 & 5 & 6 & 7 \\ 2 & 3 & 4 & 5 & 6 & 7 & 1 \end{pmatrix} = (1\ 2\ 3\ 4\ 5\ 6\ 7).$$

Just as in the $\sigma_{1,8}$ case we can easily compute the permutations corresponding to the other legal moves. Below we list the permutations corresponding to the legal moves in (2.2) other than $\sigma_{i,i+1}$, $1 \leq i \leq 15$:

$$\sigma_{1,8} = (1\ 2\ 3\ 4\ 5\ 6\ 7),$$
$$\sigma_{2,7} = (2\ 3\ 4\ 5\ 6),$$
$$\sigma_{3,6} = (3\ 4\ 5),$$
$$\sigma_{7,10} = (7\ 8\ 9),$$
$$\sigma_{6,11} = (6\ 7\ 8\ 9\ 10),$$
$$\sigma_{5,12} = (5\ 6\ 7\ 8\ 9\ 10\ 11),$$
$$\sigma_{9,16} = (9\ 10\ 11\ 12\ 13\ 14\ 15),$$
$$\sigma_{10,15} = (10\ 11\ 12\ 13\ 14),$$
$$\sigma_{11,14} = (11\ 12\ 13).$$

The permutations corresponding to the legal moves are all cycles of odd length. By Proposition 2.9.17 a cycle of odd length is an even permutation. The upshot is that if we have a configuration C_1 and perform a series of legal moves corresponding to permutations $\tau_1, \ldots, \tau_n$ and finally reaching the configuration C_2 then

$$\sigma_{C_2}^{-1} = \tau_n \cdots \tau_1 \sigma_{C_1}^{-1},$$

and therefore $\mathrm{sgn}(\sigma_{C_2}) = \mathrm{sgn}(\sigma_{C_1})$. So unless two configurations map to permutations of the same sign you cannot come from one to the other through a sequence of legal moves. This proves that the original Loyd puzzle is unsolvable. This could have been verified without going through the machinery of writing legal moves as permutations in A_{15}. Using the permutation description of the legal moves we can actually prove more, as follows.

A surprising fact is that if two configurations map to permutations of the same sign then you can come from one to the other using a sequence of legal moves. We will give a simple proof of this here. First we need a small lemma. We call a 3-cycle *simple* if it has the form $(k\ k + 1\ k + 2)$.

Lemma 2.9.20 *Every 3-cycle is a product of simple 3-cycles in A_n if $n \geq 3$.*

Proof. This is proved by induction. For $n = 3$ one gets all 3-cycles as powers of the simple 3-cycle $(1\ 2\ 3)$. If $n > 3$ we may assume by induction that every 3-cycle not containing both 1 and n can be written as a product of simple 3-cycles. Consider the 3-cycle $(1\ x\ n)$ containing both 1 and n. Choose $y \notin \{1, x, n\}$. Then $(1\ x\ n) = (1\ x\ y)(x\ n\ y)$ and $(1\ n\ x) = (1\ x\ n)^2$. This proves by induction that every 3-cycle in A_n can be written as a product of simple 3-cycles. □

Now we get by Lemma 2.9.18 that every even permutation is a product of simple 3-cycles. This leads us to the main result:

Theorem 2.9.21 *Every permutation in A_{15} is a product of permutations corresponding to legal moves in the* 15-puzzle.

Proof. It suffices to prove that all the simple 3-cycles can be written as products of the legal moves. We will show how to get the simple 3-cycles (1 2 3), ..., (5 6 7) and leave the rest to the reader. Consider the two legal moves $\tau = (1\ 2\ 3\ 4\ 5\ 6\ 7)$ and $\sigma = (3\ 4\ 5)$. Then

$$\tau\sigma\tau^{-1} = (\tau(3)\ \tau(4)\ \tau(5)) = (4\ 5\ 6)$$

by Lemma 2.9.8. Similarly $\tau^2\sigma\tau^{-2} = (5\ 6\ 7)$, $\tau^5\sigma\tau^{-5} = (1\ 2\ 3)$ and $\tau^6\sigma\tau^{-6} = (2\ 3\ 4)$. $\square$

Suppose we have two configurations C_1, C_2 for which $\mathrm{sgn}(\sigma_{C_1}) = \mathrm{sgn}(\sigma_{C_2})$. Then $\sigma_{C_2}^{-1}\sigma_{C_1} \in A_{15}$. This means that $\sigma_{C_2}^{-1}\sigma_{C_1}$ can be written as a product $\tau_1 \cdots \tau_r$ of permutations corresponding to legal moves, by Theorem 2.9.21. Thus $\sigma_{C_2}^{-1} = \tau_1 \cdots \tau_r\sigma_{C_1}^{-1}$. We can translate this back into a sequence of legal moves turning C_1 into C_2. This is done by placing the empty square in the appropriate cell according to each permutation (recall that the permutation does not change when the empty square is moved along the snake pattern). For example for $\tau = \sigma_{i,j}$ we move the empty square to cell i in order to carry out the move from cell i to cell j.

2.10 Actions of groups

Groups are very powerful algebraic objects in themselves, but most of the time it is more interesting to know how they interact with the world around them. The relevant notion is that of a group acting on a set. In this section we will apply actions of groups to combinatorics and counting, to conjugacy classes in the symmetric groups and to the proof of the Sylow theorems.

Definition 2.10.1 Let G be a group and S a set. We will say that G acts (from the left) on S if there is a map

$$\alpha : G \times S \to S,$$

denoted $\alpha(g, s) = g \cdot s$, such that

(i) $e \cdot s = s$ for every $s \in S$,
(ii) $(g \cdot h) \cdot s = g \cdot (h \cdot s)$ for every $g, h \in G$ and every $s \in S$.

When no confusion is likely to arise we will leave out the multiplication point from $g \cdot s$ and just write gs.

Definition 2.10.2   Let $\alpha : G \times S \to S$ be an action of G on S, $X \subseteq S$ a subset of S and $s \in S$ an element of S. Then $G \cdot s = Gs = \{gs \mid g \in G\}$ is called the *orbit* of s (under the action of G). The set of orbits $\{Gs \mid s \in S\}$ is denoted S/G. Let $g \cdot X = gX = \{gx \mid x \in X\}$, where $g \in G$. Then

$$G_X = \{g \in G \mid gX = X\}$$

is called the stabilizer of X. If $X = \{x\}$ we denote G_X by G_x. A fixed point for the action is an element $s \in S$ such that $gs = s$ for every $g \in G$. The set of fixed points is denoted S^G.

Example 2.10.3   The above definitions may seem abstract, but we have already seen many examples of them.

(i) The symmetric group S_n acts on the set $M_n = \{1, 2, \ldots, n\}$ in the natural way $\sigma i = \sigma(i)$, where $\sigma \in S_n$ and $i \in M_n$. The stabilizer $(S_n)_i$ consists of the permutations fixing i. Let $\sigma \in S_n$ and let H denote the subgroup $\langle \sigma \rangle$. Then we have an action $\alpha_H : H \times S \to S$ (given by $\sigma^n i = \sigma^n(i)$, where $n \in \mathbb{N}$ and $i \in S$). The orbits of this action are in one-to-one correspondence with the disjoint cycles of σ (see Proposition 2.9.6).
(ii) Let H be a subgroup of a group G. Then we have an action $\alpha : H \times G \to G$ given by

$$\alpha(h, g) = h \cdot g = gh^{-1} \text{ (why do we need } h^{-1} \text{ and not just } h\text{?) .}$$

The orbit $H \cdot g$ is the left coset gH. The set of orbits of this action is the set G/H of left cosets of H. Notice that this action does not have any fixed points.
(iii) Let L be the group of linear isometries of $\mathbb{R}^2$ (Example 2.1.12). Then there is a natural action $\alpha : L \times \mathbb{R}^2 \to \mathbb{R}^2$ given by $\alpha(\varphi, v) = \varphi(v)$. The stabilizer L_K of $K \subseteq \mathbb{R}^2$ is the group D_3, where K is the triangle in Example 2.1.13. The origin $(0, 0)$ is the only fixed point of this action. The orbit $L(x, y)$ is the circle centered in the origin with radius $\sqrt{x^2 + y^2}$.

We have singled out the following example of a group action, because it is important in almost all mathematics. We will make use of it later when proving the Sylow theorems.

Example 2.10.4 Let G be a group and H a subgroup. We may not be able to make G/H into a group (H is not necessarily normal), but there is an action $\alpha : G \times G/H \to G/H$ of G, on the left cosets of H, given by $\alpha(g, g'H) = (gg')H$ where $g, g' \in G$. This is an action with only one orbit.

Proposition 2.10.5 *Let $\alpha : G \times S \to S$ be an action.*

(i) *Let $X \subseteq S$ be a subset of S. Then G_X is a subgroup of G.*
(ii) *The set S is the union of G-orbits*

$$S = \bigcup_{s \in S} Gs,$$

   where $Gs \neq Gt$ implies $Gs \cap Gt = \emptyset$ if $s, t \in S$.
(iii) *Let $x \in S$. Then*

$$\tilde{f} : G/G_x \to Gx$$

   given by $\tilde{f}(gG_x) = gx$ is a well defined and bijective map between the left cosets of G_x and the orbit Gx.

Proof. (i) Clearly $e \in G_X$, and if $g, h \in G_X$ then $gh \in G_X$. If $g \in G_X$ then $g^{-1}X = g^{-1}(gX) = eX = X$, so that $g^{-1} \in G_X$. This shows that G_X is a subgroup of G.

(ii) If $s \in S$ then $es \in S$, so that $s \in Gs$. This shows that $S = \cup_{s \in S} Gs$. Let us prove that $Gs \neq Gt$ gives $Gs \cap Gt = \emptyset$. Suppose that $z \in Gs \cap Gt \neq \emptyset$. Then we can find $g_1, g_2 \in G$ such that $z = g_1 s = g_2 t$. This implies that $s = es = g_1^{-1}(g_1 s) = g_1^{-1}(g_2 t) = (g_1^{-1} g_2)t$, so that $s \in Gt$ and thereby $Gs \subseteq Gt$. In the same way we get that $Gt \subseteq Gs$, so that $Gs = Gt$.

(iii) Let $g_1, g_2 \in G$. Then $g_1 x = g_2 x$ if and only if $x = (g_1^{-1} g_2)x$ if and only if $g_1^{-1} g_2 \in G_x$. By Lemma 2.2.6 we get $g_1 x = g_2 x$ if and only if $g_1 G_x = g_2 G_x$. So $\tilde{f}(gG_x) = gx$ is a well defined and injective map. Since it is also surjective, it is a bijective map.   □

Example 2.10.6 Recall that the group L of linear isometries acts naturally on $\mathbb{R}^2$ via $\varphi v = \varphi(v)$, where $\varphi \in L$ and $v \in \mathbb{R}^2$. Suppose that $O \subseteq \mathbb{R}^2$ is

an octagon

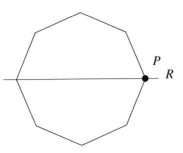

centered at $(0, 0)$. We consider the set G of linear isometries mapping O to itself:

$$G = \{\varphi \in L \mid \varphi(O) = O\}.$$

In the language of group actions G is the stabilizer of O, i.e. $G = L_O$. How do we determine the order $|G|$ of G? Let us first see that $|G|$ is finite. The following argument is quite general and can be used in many other circumstances. Let the vertices of O be listed as $V = \{1, 2, \ldots, 8\}$. If $g \in G$ and $v \in V$ then $gv \in V$, since g is a reflection or a rotation by Example 2.1.12. This shows that G acts on V and that we have a group homomorphism $\varphi : G \to S_8$ given by $\varphi(g)(v) = gv$. If $gv = v$ for every $v \in V$ then g must be the identity, since g is a linear map fixing a basis of $\mathbb{R}^2$. Therefore φ is injective and $|G| \leq 8!$. The orbit of the vertex P is $GP = V$. Now we can use the formula $|G/G_P| = |GP| = |V|$ from Proposition 2.10.5(iii) to compute the order of G. The stabilizer G_P consists only of the identity and the reflection in the line R. So $|G_P| = 2$. Therefore $|V| = |G/G_P| = |G|/2$ and thus $|G| = 2 \cdot 8 = 16$.

The method illustrated in Example 2.10.6 becomes even more useful when computing orders of "symmetry" groups in $\mathbb{R}^3$ (such as stabilizers of the cube or the regular dodecahedron under the action of the group SO_3 of rotations of $\mathbb{R}^3$).

From Proposition 2.10.5 we can also deduce the following important counting formula.

Corollary 2.10.7 *Let $G \times S \to S$ be an action, where S is a finite set. Then*

$$|S| = |S^G| + \sum_x |G/G_x|,$$

where the summation is done by picking out an element x from each orbit with more than one element.

Proof. By Proposition 2.10.5(ii), we may count the number of elements in S by counting the number of elements in each orbit and adding these. The formula expresses this, in that we first count the orbits containing one point (this is the term $|S^G|$) and then the orbits containing two or more points (this is the summation). In the latter case we use the bijection $Gx \to G/G_x$ from Proposition 2.10.5(iii). □

The following lemma is a very valuable tool for doing combinatorics and counting. Notice how having two different ways of counting leads to a surprising formula.

Lemma 2.10.8 (Burnside) *Let $G \times S \to S$ be an action, where G is a finite group and S a finite set. Then*

$$|S/G| = \frac{\sum_{g \in G} |S^g|}{|G|}$$

where $S^g = \{x \in S \mid gx = x\}$.

Proof. Define $T = \{(g, x) \in G \times S \mid gx = x\}$. We will count the elements in T in two different ways. For every $g \in G$ we count the number of $x \in S$ fixed by g. This is the same as for every $x \in S$ counting the number of $g \in G$ that fixes x. Thus we have the formula

$$|T| = \sum_{g \in G} |S^g| = \sum_{x \in S} |G_x|.$$

The last sum can be rewritten using Proposition 2.10.5(ii):

$$\sum_{x \in S} |G_x| = \sum_{\text{orbits } Gx} \sum_{y \in Gx} |G_y| = \sum_{\text{orbits } Gx} |Gx||G_x| = \sum_{\text{orbits } Gx} |G| = |S/G| |G|,$$

since $Gy = Gx$ when $y \in Gx$ and therefore $|G_y| = |G_x|$, by Proposition 2.10.5(iii). This gives the desired result. □

Example 2.10.9 Suppose that you color four of the edges of the octagon in Example 2.10.6 white and four black. You can do this in $\binom{8}{4} = 70$ ways, but some of them can be mapped to each other using reflections and rotations. We wish to count the number of essentially different colorings. The group G from Example 2.10.6 acts on the set S of colorings without taking into account that some of them are the same. So $|S| = 70$. The colorings in the same orbits of this action are considered as the same (two colorings are in the same orbit if you can reflect or rotate one to the other).

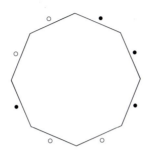

We wish to find the number of orbits $|S/G|$ using Lemma 2.10.8. Now, G has 16 elements consisting of eight reflections and eight rotations. We need to find $|S^g|$ for $g \in G$. Let g be a reflection in a line through two opposite vertices of the octagon. Once we have chosen the colors of two edges of the four on one side of the line, the colors of the rest of the edges are determined if the coloring is invariant under g. This means that $|S^g| = \binom{4}{2} = 6$. Now let g be a reflection in a line through the midpoints of opposite edges. The color of these two opposite edges has to be the same for the coloring to be invariant under g. Therefore $|S^g| = 2 \cdot 3 = 6$.

Of course, $|S^e| = 70$. If g is a rotation of $\pi/4$, $3\pi/4$, $5\pi/4$ or $7\pi/4$ then $|S^g| = 0$. If g is a rotation of $\pi/2$ or $3\pi/2$ then $|S^g| = 2$. Finally, if g is a rotation of π then $|S^g| = \binom{4}{2} = 6$. Plugging these numbers into Burnside's formula gives

$$|S/G| = \frac{1}{16}(6 + 6 + 6 + 6 + 6 + 6 + 6 + 6 + 70 + 2 + 2 + 6) = 8.$$

Example 2.10.10   Let us look again at Example 2.10.9 but now consider only the group G of rotations acting on the set S of colorings. So G consists of rotations of $2k\pi/8$, where $k = 0, 1, \ldots, 7$. Two colorings are considered the same if they are in the same orbit under the action of G. This means that you can map one to the other using a rotation in G. How many essentially different colorings are there now? Again this amounts to counting the number of orbits of G in S. We already have the relevant numbers $|S^g|$ for $g \in G$ from Example 2.10.9. Let us plug them into Burnside's formula and compute

$$|S/G| = \frac{1}{8}(70 + 0 + 2 + 0 + 6 + 0 + 2 + 0) = 10.$$

2.10.1 Conjugacy classes

The map $\alpha : G \times G \to G$ given by $\alpha(g, h) = ghg^{-1}$ is an action of G on G. It is called *conjugation*. The orbit

$$G \cdot h = C(h) = \{ghg^{-1} \mid g \in G\}$$

is denoted $C(h)$ and called the *conjugacy class* containing h. The stabilizer G_h is denoted $Z(h)$ and called the *centralizer* of h. Notice that

$$Z(h) = \{g \in G \mid gh = hg\}.$$

The set of fixed points

$$G^G = Z(G) = \{g \in G \mid gx = xg \text{ for every } x \in G\}$$

is denoted $Z(G)$ and called the *center* of G. There is at least one fixed point for the conjugation action, namely $e \in Z(G)$. In fact $Z(G)$ is an abelian normal subgroup of G (see Exercise 2.50). The stabilizer of a subgroup $H \subseteq G$,

$$G_H = N_G(H) = \{g \in G \mid gHg^{-1} = H\},$$

is denoted $N_G(H)$ and called the *normalizer* of H in G. Notice that H is a normal subgroup if and only if $N_G(H) = G$ (see Exercise 2.51). If G is a finite group then we may write Corollary 2.10.7 as

$$|G| = |Z(G)| + \sum_{h \in G} |G/Z(h)|,$$

where the last sum is done by picking out one element h from each conjugacy class with more than one element.

2.10.2 Conjugacy classes in the symmetric group

Conjugacy classes in the symmetric group S_n have a very appealing description. Let $\sigma \in S_n$ and write $\sigma = \sigma_1 \sigma_2 \cdots \sigma_r$ as a product of disjoint cycles (Proposition 2.9.6) of increasing length $i_1 \leq i_2 \leq \cdots \leq i_r$. We get for example that

$$(3\,4) = (1)(2)(3\,4)$$

for $(3\,4) \in S_4$, so that $i_1 = i_2 = 1$ and $i_3 = 2$. The increasing sequence $i_1 \leq \cdots \leq i_r$ is called the *cycle type* of σ. It follows by Lemma 2.9.8 that the conjugacy class $C(\sigma)$ consists of permutations with the same cycle type as σ. You may see this by writing on top of each other two permutations $\sigma_1, \sigma_2 \in S_n$ with the same cycle type. Let $\tau \in S_n$ be given by mapping the elements on top to the elements below. Then $\tau \sigma_1 \tau^{-1} = \sigma_2$. An example will clarify this.

Example 2.10.11 Let $\sigma_1 = (1\,2)(3\,4)$ and $\sigma_2 = (2\,3)(4\,1)$ be permutations in S_4. Then we can write σ_1 and σ_2 on top of each other as follows:

$$(1\,2)(3\,4)$$
$$(2\,3)(4\,1).$$

Now define τ by

$$\begin{pmatrix} 1 & 2 & 3 & 4 \\ 2 & 3 & 4 & 1 \end{pmatrix}.$$

Then we see that $\tau \sigma_1 \tau^{-1}$ is given by

$$\tau(1\,2)(3\,4)\tau^{-1} = \tau(1\,2)\tau^{-1}\tau(3\,4)\tau^{-1} = (\tau(1)\,\tau(2))(\tau(3)\,\tau(4)) = \sigma_2.$$

Counting the number of elements in a conjugacy class is a combinatorial problem. The number of conjugacy classes in S_n is the number of sequences $1 \le i_1 \le i_2 \le \cdots \le i_r$ of integers with

$$i_1 + \cdots + i_r = n.$$

For example, there are five conjugacy classes in S_4, corresponding to the sequences

$$1 \le 1 \le 1 \le 1,$$
$$1 \le 1 \le 2,$$
$$1 \le 3,$$
$$2 \le 2,$$
$$4.$$

The permutation $\sigma = (3\,4)$ lies in the conjugacy class corresponding to $1 \le 1 \le 2$. The conjugacy class $C(\sigma)$ consists of the elements

$$\{(3\,4), (2\,4), (2\,3), (1\,4), (1\,3), (1\,2)\}.$$

Therefore $|Z(\sigma)| = |S_4|/|C(\sigma)| = 4$.

Remark 2.10.12 Counting the number of conjugacy classes in S_n translates into the problem of counting increasing sequences $1 \le i_1 \le i_2 \le \cdots \le i_r \le n$ such that $i_1 + \cdots + i_r = n$. For example, when $n = 6$ there are the following increasing sequences:

$$6 = 1 + 1 + 1 + 1 + 1 + 1,$$
$$6 = 1 + 1 + 1 + 1 + 2,$$
$$6 = 1 + 1 + 2 + 2,$$
$$6 = 2 + 2 + 2,$$
$$6 = 1 + 1 + 1 + 3,$$
$$6 = 1 + 2 + 3,$$
$$6 = 3 + 3,$$
$$6 = 1 + 1 + 4,$$
$$6 = 2 + 4,$$
$$6 = 1 + 5,$$
$$6 = 6.$$

This combinatorial problem was studied by Euler in his landmark work *Introductio in Analysin Infinitorum* (1748). Let $p(n)$ be the number of ways in which an integer n can be written as a sum of natural positive numbers. Note that $p(0) = 1$, counting the empty sum as a way of writing 0, and $p(n) = 0$ if $n < 0$. We have seen above that $p(6) = 11$. Euler proved the remarkable identity

$$p(n) = p(n-1) + p(n-2) - p(n-5) - p(n-7) \qquad (2.3)$$
$$+ p(n-12) + p(n-15) - \cdots$$

where the numbers subtracted from n are $\frac{1}{2}(3k^2 \mp k)$, $k = 1, 2, \ldots$.

2.10.3   Groups of order p^r

A finite group of order p^r, where p is a prime number and $r \in \mathbb{N}$, is called a *p-group*.

Proposition 2.10.13 *Let G be a non-trivial p-group acting on a finite set S. Then $|S| \equiv |S^G| \pmod{p}$.*

Proof.   Corollary 2.10.7 gives

$$|S| = |S^G| + \sum_{x \in S} |G/G_x|,$$

where the summation on the right hand side is done by picking out an element x from each orbit with more than one element (x is not a fixed point). If $x \in S$ is not a fixed point then G_x is a proper subset of G. Therefore p divides $|G/G_x| = |G|/|G_x|$. Thus p divides every term in the summation on the right hand side. Therefore p divides $|S| - |S^G|$. $\square$

Corollary 2.10.14 *Let G be a non-trivial p-group of order p^r. Then*

$$|G| \equiv |Z(G)| \,(\mathrm{mod}\; p)$$

and $|Z(G)| > 1$.

Proof. This is done simply by using Proposition 2.10.13 for the conjugation action. In this case the $Z(G)$ are the fixed points. Since $p \nmid p^r - 1$ we obtain $|Z(G)| > 1$. $\square$

Corollary 2.10.15 *Let p be a prime number. A group G of order $|G| = p^2$ is abelian.*

Proof. We will prove that $|Z(G)| = |G|$. By Corollary 2.10.14, we get $|Z(G)| > 1$. Since $|Z(G)|$ divides $|G|$, the only possibilities left are $|Z(G)| = p$ or $|Z(G)| = p^2$. We wish to exclude $|Z(G)| = p$. Suppose that this is the case. Since $Z(G) \subseteq G$ is a normal subgroup, $G/Z(G)$ is a group of order p. Therefore it has to be cyclic, by Proposition 2.7.2. Let $xZ(G)$ be a generator for $G/Z(G)$, where $x \in G$. Then every $gZ(G) = x^n Z(G)$ for some power $n \in \mathbb{N}$. In particular every element $g \in G$ can be written $g = x^n a$, where $a \in Z(G)$. But $(x^m a)(x^n b) = (x^n b)(x^m a)$ when $a, b \in Z(G)$. This proves that G is abelian, contradicting $|Z(G)| = p < |G|$. $\square$

Extending the method in the proof of Corollary 2.10.15 a little, one can show that $\mathbb{Z}/p^2\mathbb{Z}$ and $\mathbb{Z}/p\mathbb{Z} \times \mathbb{Z}/p\mathbb{Z}$ are the only groups of order p^2 up to isomorphism. There is also a small modification to the proof that makes it simpler: if $|Z(G)| = p$ then there must exist $g \in G \setminus Z(G)$. But then $Z(G) \subsetneq Z(g)$. This implies that $Z(g) = G$ or that $g \in Z(G)$, which is a contradiction.

2.10.4 The Sylow theorems

We now move on to the celebrated Sylow theorems. Sylow (1832–1918) published a 10-page paper [24] in *Mathematische Annalen* in 1872 containing three theorems. His three theorems have survived to the present day and are of fundamental importance.

Definition 2.10.16 Let G be a finite group and p a prime number, and suppose that $|G| = p^r m$, where $p \nmid m$. A *Sylow p-subgroup* is a subgroup $H \subseteq G$ of order p^r.

Theorem 2.10.17 (First Sylow theorem) *Let G be a finite group and p a prime number, and suppose that $|G| = p^r m$, where $p \nmid m$. Then G contains a Sylow p-subgroup.*

Proof. Define a map $\alpha : G \times S \to S$, where $S = \{X \subseteq G \mid |X| = p^r\}$, given by $\alpha(g, X) = \{gx \mid x \in X\}$, where $X \in S$ and $g \in G$. It follows from subsection 2.1.6 that $\alpha(g, X) \in S$ when $g \in G$ and $X \in S$. It is easy to see that α is an action of G on S. The number of subsets with p^r elements in a set having $p^r m$ elements is given by the binomial coefficient

$$|S| = \binom{p^r m}{p^r} = \frac{p^r m (p^r m - 1) \cdots (p^r m - p^r + 1)}{p^r (p^r - 1) \cdots 1}.$$

Since $p^{r+1} \nmid p^r m - i$ and $p^{r+1} \nmid p^r - i$ for $i = 0, \ldots, p^r - 1$, it follows that the highest power of p dividing $p^r - i$, is the highest power of p dividing $p^r m - i$, for $i = 0, \ldots, p^r - 1$. From this we deduce the important fact that $p \nmid |S|$.

By Proposition 2.10.5(ii), there must exist an orbit $G \cdot X$, where $X \in S$, such that $p \nmid |G \cdot X|$. From $|G_X||G \cdot X| = |G|$ (Proposition 2.10.5(iii)), it follows that p^r divides $|G_X|$. We will show that $|G_X| = p^r$. To this end we use the action of G_X on X itself. The orbits of this action are the right cosets $G_X g$ of G_X. So the orbits each have $|G_X|$ elements (why?). Again by Proposition 2.10.5(ii) we get that $|G_X|$ divides $|X| = p^r$. This proves that $|G_X| = p^r$ and therefore that G_X is a Sylow p-subgroup of G.   □

Theorem 2.10.18 (Second Sylow theorem) *Let G be a finite group and P, Q two Sylow p-subgroups. Then there exists $g \in G$ such that*

$$g P g^{-1} = Q.$$

Furthermore, any p-subgroup H is contained in a Sylow p-subgroup.

Proof. The natural action of G on G/Q (Example 2.10.4) restricts to give an action of P on G/Q. Since p does not divide $|G/Q| = |G|/|Q|$, this action has a fixed point by Proposition 2.10.13. Thus we can find a left coset xQ, $x \in G$, such that $gxQ = xQ$ for every $g \in P$. This means that $P \subseteq xQx^{-1}$. But since $|P| = |Q| = |xQx^{-1}|$, we must have $P = xQx^{-1}$.

Let H be a non-trivial p-subgroup and P a Sylow p-subgroup. As above H acts on G/P and has a fixed point yP, $y \in G$. This means that $hyP = yP$ for every $h \in H$ and therefore that $H \subseteq yPy^{-1}$, so that H is contained in the Sylow p-subgroup yPy^{-1}.   □

Theorem 2.10.19 (Third Sylow theorem)   *Let G denote a finite group of order $p^r m$, where $p \nmid m$. Let $\mathrm{Syl}_p(G)$ denote the set of Sylow p-subgroups. Then*

(i) $|\mathrm{Syl}_p(G)|$ *divides m,*
(ii) $|\mathrm{Syl}_p(G)| \equiv 1 \,(\mathrm{mod}\ p)$.

Proof.   Let P be a Sylow p-subgroup. Then G acts on $\mathrm{Syl}_p(G)$ by conjugation. This action has only one orbit, by the second Sylow theorem. Thus by Proposition 2.10.5(iii) we get

$$|\mathrm{Syl}_p(G)| = |G/N_G(P)|,$$

where P is a Sylow p-subgroup. But since $P \subseteq N_G(P)$, it follows that $|G/P| = |G/N_G(P)|\,|N_G(P)/P|$ (see Exercise 2.18). Therefore $|\mathrm{Syl}_p(G)|$ divides $|G/P| = m$. This proves (i).

The conjugation action of G on $\mathrm{Syl}_p(G)$ restricts to give an action of P on $\mathrm{Syl}_p(G)$. To prove (ii), it suffices by Proposition 2.10.13 to show that P is the only fixed point for this action. Suppose that $Q \in \mathrm{Syl}_p(G)$ is a fixed point, i.e. $gQg^{-1} = Q$ for every $g \in P$. This means that $P \subseteq N_G(Q)$. Now using the second Sylow theorem on the Sylow p-subgroups P and Q of the group $N_G(Q)$, there must exist $g \in N_G(Q)$ such that $Q = gQg^{-1} = P$. This shows that P is the only fixed point.   □

A typical example of the use of the Sylow theorems is the following (more examples are found in Exercises 2.52–2.56).

Example 2.10.20   A group G of order 143 must be isomorphic to $\mathbb{Z}/143\mathbb{Z}$. Since $143 = 11 \cdot 13$, the third Sylow theorem tells us that

$$|\mathrm{Syl}_{11}(G)| \in \{1, 13\}, \quad |\mathrm{Syl}_{13}(G)| \in \{1, 11\}$$

and

$$|\mathrm{Syl}_{11}(G)| \equiv 1 \,(\mathrm{mod}\ 11), \quad |\mathrm{Syl}_{13}(G)| \equiv 1 \,(\mathrm{mod}\ 13).$$

So there is a unique Sylow 11-subgroup P and a unique Sylow 13-subgroup Q in G. These Sylow subgroups have to be normal (why?). The product PQ is a subgroup (see Lemma 2.3.6) and it contains P and Q. This implies that $PQ = G$, as $11 = |P|$ divides $|PQ|$ and $13 = |Q|$ divides $|PQ|$. Since $P \cap Q$ is a proper subgroup of Q, it follows that $P \cap Q = \{e\}$ by Theorem 2.2.8. This implies by Lemma 2.8.1 that

$$\pi : P \times Q \xrightarrow{\sim} G$$

given by $\pi(p, q) = pq$ is an isomorphism. So G is isomorphic to $\mathbb{Z}/11\mathbb{Z} \times \mathbb{Z}/13\mathbb{Z}$, which by the Chinese remainder theorem (Proposition 2.8.2) is isomorphic to $\mathbb{Z}/143\mathbb{Z}$.

2.11 Exercises

1. Let G be a group and $g \in G$ an element of G. Prove that the map $\xi : G \to G$ given by $\xi(x) = xg$ is bijective.
2. Using subsection 2.1.6 construct the possible composition tables for a group with four elements.
3. Verify the composition table in Example 2.1.6.
4. Let G be a group and $H \subseteq G$ a non-empty subset. Prove that H is a subgroup if and only if $xy^{-1} \in H$ for all $x, y \in H$.
5. Let H be a non-empty finite subset of a group G. Prove that H is a subgroup if $xy \in H$ for every $x, y \in H$. Give an example where this breaks down if H is infinite. (Hint: consider $e, x, x^2, \ldots$ or use the fact that multiplication by $x \in H$ is bijective.)
6. Prove in detail that $GL_2(\mathbb{R})$ and $O_2(\mathbb{R})$ are groups and that they are non-abelian.
7. In the notation of Example 2.1.6, show that $\{e, d, f\}$ is a normal subgroup of S_3. List the subgroups of order 2. Are any of these normal?
8. Let $\varphi : \mathbb{R}^2 \to \mathbb{R}^2$ be an isometry such that $\varphi((0, 0)) = (0, 0)$, as in Example 2.1.12.
   (i) Prove that $\varphi(v_1) \cdot \varphi(v_2) = v_1 \cdot v_2$, where $v_1, v_2 \in \mathbb{R}^2$ and $\cdot$ denotes the usual inner product on $\mathbb{R}^2$ (use $|v_1 - v_2|^2 = |v_1|^2 + |v_2|^2 - 2v_1 \cdot v_2$).
   (ii) Show that $\varphi(e_1) \cdot \varphi(e_2) = 0$ and that $\varphi(\lambda e_1 + \mu e_2) = \lambda \varphi(e_1) + \mu \varphi(e_2)$, where e_1, e_2 is the usual basis of $\mathbb{R}^2$ and $\lambda, \mu \in \mathbb{R}$.
   (iii) Prove that φ is a homomorphism (linear map) of vector spaces, i.e.

(a) $\varphi(\lambda v) = \lambda\varphi(v)$, where $\lambda \in \mathbb{R}$ and $v \in \mathbb{R}^2$,

(b) $\varphi(v_1 + v_2) = \varphi(v_1) + \varphi(v_2)$, where $v_1, v_2 \in \mathbb{R}^2$.

(iv) Prove that φ is invertible by proving that its determinant is non-zero.

9. Let L denote the group of linear isometries (rotations and reflections) of $\mathbb{R}^2$ (see Example 2.1.12). Consider the square $K \subseteq \mathbb{R}^2$.

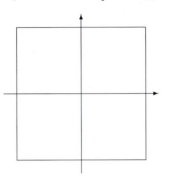

(i) List the elements of the group $G = \{\varphi \in L \mid \varphi(K) = K\}$.

(ii) Write down the composition table for G.

10. Write down the subgroups of $\mathbb{Z}/6\mathbb{Z}$.

11. Why are $\{[0]\}$ and $\mathbb{Z}/7\mathbb{Z}$ the only subgroups of $\mathbb{Z}/7\mathbb{Z}$?

12. Show that a group G is not the union of two proper subgroups $H_1, H_2 \subsetneq G$. Can a group be the union of three proper subgroups?

13. Let N be a normal subgroup of a group G. Prove that $gN = Ng$ for every $g \in G$.

14. Show that every subgroup of an abelian group is normal.

15. Let H be a subgroup of the group G.

(i) Show that H is a right coset and that distinct right cosets of H are disjoint.

(ii) Show that the map $\Phi : G/H \to H\backslash G$ given by $\Phi(gH) = Hg^{-1}$ is well defined. Prove also that it is bijective.

(iii) Prove that if H has index 2 in G (i.e. $|G/H| = 2$), then H is normal. Give an example of a subgroup of index 3 that is not normal.

16. Consider the subset H of $\mathrm{GL}_2(\mathbb{C})$ consisting of the eight matrices $\pm\mathbf{1}, \pm\mathbf{i}, \pm\mathbf{j}$ and $\pm\mathbf{k}$, where

$$\mathbf{1} = \begin{pmatrix} 1 & 0 \\ 0 & 1 \end{pmatrix}, \quad \mathbf{i} = \begin{pmatrix} i & 0 \\ 0 & -i \end{pmatrix}, \quad \mathbf{j} = \begin{pmatrix} 0 & 1 \\ -1 & 0 \end{pmatrix}, \quad \mathbf{k} = \begin{pmatrix} 0 & i \\ i & 0 \end{pmatrix}.$$

Verify that H is a subgroup by constructing the composition table. This group is called the *quaternion group*.

17. Prove that the quaternion group H from Exercise 2.16 is not abelian, but that all its subgroups are normal.

18. Let G be a finite group and $H \supseteq K$ subgroups of G. Prove that $|G/K| = |G/H| |H/K|$.

19. (i) Compute the inverse of $[3]$ in $(\mathbb{Z}/8\mathbb{Z})^*$.
    (ii) Compute the inverse of $[5]$ in $(\mathbb{Z}/13\mathbb{Z})^*$.

20. Prove that the inverse map of a group isomorphism is also a group homomorphism.

21. Prove that G is abelian if and only if the map $f : G \to G$ given by $f(g) = g^2$ is a group homomorphism.

22. Prove that the exponential function $\xi(x) = e^x$ is a group isomorphism from $(\mathbb{R}, +)$ to $(\mathbb{R}_{>0}, \cdot)$.

23. Using the notation of Example 2.1.6, prove that the map $\mathrm{sgn} : S_3 \to \mathbb{Z}/2\mathbb{Z}$, mapping e, d, f to $[0]$ and a, b, c to $[1]$, is a group homomorphism.

24. Prove that

$$\begin{pmatrix} 1 & 1 \\ 0 & 1 \end{pmatrix} \in \mathrm{GL}_2(\mathbb{R})$$

has infinite order in the group $\mathrm{GL}_2(\mathbb{R})$.

25. Let V be a real vector space and W a subspace of V. Show that V is an abelian group with respect to $+$ and that W is a normal subgroup in V. Prove that the quotient group V/W is a real vector space with scalar multiplication $\lambda(v + W) = \lambda v + W$, where $\lambda \in \mathbb{R}$.

26. Let G be an abelian group, K a group and $f : G \to K$ a group homomorphism. Prove that $f(G) \subseteq K$ is an abelian subgroup of K.

27. Let $\mathrm{SL}_2(\mathbb{R})$ be the subset of $\mathrm{GL}_2(\mathbb{R})$ (see Example 2.1.10) consisting of matrices with determinant 1. Show that $\mathrm{SL}_2(\mathbb{R})$ is a normal subgroup of $\mathrm{GL}_2(\mathbb{R})$. Use the isomorphism theorem to determine the group

$$\mathrm{GL}_2(\mathbb{R})/\mathrm{SL}_2(\mathbb{R}).$$

28. Prove that $(\mathbb{Z}/13\mathbb{Z})^*$ is a cyclic group by finding a generator.

29. Let p be a prime number and suppose that q is a prime number such that $q \mid 2^p - 1$. Prove that $q > p$ (hint: consider the element $[2] \in (\mathbb{Z}/q\mathbb{Z})^*$). Use this to prove that there are infinitely many prime numbers.

30. Let $\pi : G \to G/N$ be the canonical group homomorphism where N is a normal subgroup of G.
    (i) Prove that $\pi(K)$ is a subgroup of G/N if K is a subgroup of G.

(ii) Prove that $\pi^{-1}(H)$ is a subgroup of G containing N if H is a subgroup of G/N.

(iii) Prove that $\pi(\pi^{-1}(H)) = H$ and $\pi^{-1}(\pi(K)) = K$, where H is a subgroup of G/N and K is a subgroup of G containing N.

(iv) Let G be a cyclic group and $f : G \to K$ a surjective group homomorphism. Prove that K is a cyclic group.

(v) Let $N \in \mathbb{N}$. Prove using the canonical group homomorphism

$$\pi : \mathbb{Z} \to \mathbb{Z}/N\mathbb{Z}$$

that a subgroup H of $\mathbb{Z}/N\mathbb{Z}$ is cyclic.

31. (i) Write down all the elements of order 7 in $\mathbb{Z}/28\mathbb{Z}$.

   (ii) How many subgroups are there of order 7 in $\mathbb{Z}/28\mathbb{Z}$?

32. (i) Prove that the cyclic group $\mathbb{Z}/15\mathbb{Z}$ is isomorphic to the product group $\mathbb{Z}/3 \times \mathbb{Z}/5\mathbb{Z}$.

   (ii) Prove that the group $(\mathbb{Z}/15\mathbb{Z})^*$ is isomorphic to the product group $\mathbb{Z}/2 \times \mathbb{Z}/4\mathbb{Z}$. Conclude that $(\mathbb{Z}/15\mathbb{Z})^*$ is not cyclic.

33. Consider $\mathbb{Z} \subset \mathbb{Q}$ as abelian groups with $+$ as composition. Let $[q] = q + \mathbb{Z} \in \mathbb{Q}/\mathbb{Z}$, where $q \in \mathbb{Q}$.

   (i) Show that $\left[\dfrac{9}{4} \right]$ has order 4 in $\mathbb{Q}/\mathbb{Z}$.

   (ii) Determine the order of $\left[\dfrac{a}{b} \right]$ in $\mathbb{Q}/\mathbb{Z}$, where $a \in \mathbb{Z}$, $b \in \mathbb{N} \setminus \{0\}$ and $\gcd(a, b) = 1$. Conclude that every element in $\mathbb{Q}/\mathbb{Z}$ has finite order and that there are elements in $\mathbb{Q}/\mathbb{Z}$ of arbitrary large order.

   (iii) Show that $\mathbb{Q}/\mathbb{Z}$ is an infinite group that is not cyclic.

34. Prove that $(\mathbb{Q} \setminus \{0\}, \cdot)$ is not a cyclic group.

35. Give an example of a non-cyclic group of order 8.

36. Let G be a finite group of order N. Let $\psi(d)$ be the number of elements in G of order d.

   (i) Prove that $\psi(d) = 0$ if $d \nmid N$ and that G is cyclic if and only if $\psi(N) > 0$.

   (ii) Prove that

$$\sum_{d|N} \psi(d) = N.$$

   (iii) Suppose that for every divisor d in N, there is a unique subgroup H in G of order d. Prove that $\psi(d) \leq \varphi(d)$ and that G is a cyclic group.

37. Prove that an even permutation cannot be the product of an odd number of transpositions.

38. Prove that the order of a k-cycle in S_n is k.

39. Let $\tau \in S_3$ denote the 3-cycle

$$(1\,2\,3) = \begin{pmatrix} 1 & 2 & 3 \\ 2 & 3 & 1 \end{pmatrix}.$$

Show that the subgroup $\langle \tau \rangle = \{\tau^n \mid n \in \mathbb{Z}\}$ is normal in S_3.

40. Let $\sigma \in S_5$ denote the 5-cycle

$$(1\,2\,3\,4\,5) = \begin{pmatrix} 1 & 2 & 3 & 4 & 5 \\ 2 & 3 & 4 & 5 & 1 \end{pmatrix}.$$

(i) Show that σ is an even permutation and that $\langle \sigma \rangle = \{\sigma^n \mid n \in \mathbb{Z}\}$ has order 5 and write down the elements in $\langle \sigma \rangle$.

(ii) Prove that $\langle \sigma \rangle$ is not a normal subgroup of S_5.

41. (i) Let $\sigma, \tau \in S_4$. Show that $\mathrm{sgn}(\tau \sigma \tau^{-1}) = \mathrm{sgn}(\sigma)$.

(ii) Write the 3-cycle $(1\,2\,3)$ as a product of two simple transpositions. Prove that for a general 3-cycle σ one can find a permutation $\tau \in S_4$ such that $\tau \sigma \tau^{-1} = (1\,2\,3)$. Use this to show that 3-cycles in S_4 are even. Prove that a 3-cycle has order 3 in A_4.

(iii) Show that the number of 3-cycles in A_4 is greater than six. Conclude that the only subgroup of A_4 containing every 3-cycle is A_4.

(iv) Let $\varphi : A_4 \to \mathbb{Z}/2\mathbb{Z}$ be a group homomorphism. Show that if σ is a 3-cycle then $\varphi(\sigma) = [0] = 2\mathbb{Z} \in \mathbb{Z}/2\mathbb{Z}$. Use this to prove that $\varphi(\sigma) = [0]$ for every $\sigma \in A_4$.

(v) Prove that A_4 does not contain a subgroup of order 6.

42. If you are more familiar with 3-cycles this is an easier way of doing Exercise 2.41. Prove that A_n does not contain a subgroup H of index 2 (hint: consider A_n/H and deduce that H must contain all 3-cycles).

43. Write

$$\sigma = \begin{pmatrix} 1 & 2 & 3 & 4 & 5 & 6 \\ 6 & 5 & 4 & 3 & 2 & 1 \end{pmatrix} \in S_6$$

as a product of the minimal number of simple transpositions.

44. Prove that there are 45 elements of order 2 in A_6.

45. Prove that A_3 is a simple group. Prove that A_4 is not simple by proving that the elements of order 2 along with the neutral element form a normal subgroup.

46. Let K be the equilateral triangle from Example 2.1.13. Suppose that you color each edge of K using k colors. Show that the number of

colorings is

$$\frac{1}{6}(k^3 + 3k^2 + 2k),$$

where two colorings are considered the same if they map to each other using rotations and reflections.

47. (i) Give a coloring from each orbit in Example 2.10.9.
    (ii) Give a coloring from each orbit in Example 2.10.10.
    (iii) Comparing Example 2.10.9 with Example 2.10.10, which colorings are invariant under rotations but not under reflections?

48. In how many ways can you color the 16 squares of a 4×4 board when half of them must be black and the other half white? Now answer the same question when colorings are considered the same if they map to each other using rotations and reflections.

49. Consider the permutations $\sigma_1 = (1)(2)(345)$, $\sigma_2 = (3)(4)(152)$ and $\tau = (13)(245)$ in S_5.
    (i) What is the minimal number of simple transpositions needed in writing τ as a product of simple transpositions?
    (ii) Show that $\tau \notin A_5$ and that

    $$\tau \sigma_1 \tau^{-1} = \sigma_2.$$

    (iii) Show that $\sigma_1, \sigma_2 \in A_5$, $\tau_1 = (34)\tau \in A_5$ and $\tau_1 \sigma_1 \tau_1^{-1} = \sigma_2$.
    (iv) Now we know that σ_1, σ_2 are conjugate via a permutation τ_1 in A_5. Show that a permutation of the cycle type (a) $1 \le 1 \le 1 \le 1 \le 1$, (b) $1 \le 2 \le 2$, (c) $1 \le 1 \le 3$ or (d) 5 is even. We know that permutations of the same cycle type are conjugate via a permutation in S_5. Show that two permutations with the same cycle type, (a), (b) or (c), are conjugate via a permutation in A_5.
    (v) Give an example of two 5-cycles that cannot be conjugate via a permutation in A_5.
    (vi) Show that in general a normal subgroup N in a group G is a disjoint union of conjugacy classes $C(n)$, $n \in N$ (subsection 2.10.1).
    (vii) One may prove on further inspection that A_5 is the disjoint union of conjugacy classes with 1, 12, 12, 15 and 20 elements (check that $1 + 12 + 12 + 15 + 20 = 60$). Thus show that A_5 is a simple group.

50. Let G be a group. Prove that the center $Z(G)$ of the group is an abelian normal subgroup of G.

51. Let $H \subseteq G$ be a subgroup of a group G. Prove that the normalizer $N_G(H)$ is a subgroup of G containing H. Prove that H is normal if and only if $G = N_G(H)$.

52. Let G be a finite group and p a prime number. Prove that G contains an element of order p if p divides $|G|$. (Hint: reduce to the situation where G is cyclic and of order p^r).

53. Prove that a group of order 15 is cyclic.

54. Does a group of order 14 have to be cyclic?

55. Compute the number of elements of order 5 in a group of order 20.

56. Let p and q be prime numbers. Prove that a finite group G of order pq cannot be simple.

57. **(HOF)** Prove using only ideas developed in Chapter 2 that a finite abelian group is isomorphic to a product of cyclic groups.

3 Rings

A ring is an abelian group with a multiplication. The situation is very similar to the integers $\mathbb{Z}$. We know that $(\mathbb{Z}, +)$ is an abelian group, but at the same time we have multiplication as an additional composition. Rings were introduced by the German mathematician R. Dedekind (1831–1916), a student of Gauss, in connection with his studies of algebraic numbers, complex numbers that are roots of polynomials with integer coefficients. The definition of a ring appears in Dedekind's supplements to Dirichlet's book *Zahlentheorie* in the late nineteenth century. The theory of rings forms a wide framework useful in solving equations, computing with congruences, solving problems in number theory and exploring quantum mathematics. We will mostly deal with commutative rings (such as $\mathbb{Z}$), for which factors can be interchanged.

Ideals are certain subgroups of commutative rings that satisfy one crucial property producing new (quotient) rings, just as normal subgroups give rise to new (quotient) groups. Ideals were originally born out of failed, but very clever, attempts to prove Fermat's last theorem. In order to understand the definitions and concepts of this chapter it is advisable to be extremely concrete. Each time you encounter a new definition or a new concept check it with your examples. The main examples in this chapter are the integers $\mathbb{Z}$, finite quotient rings $\mathbb{Z}/n\mathbb{Z}$ of the integers, the Gaussian integers $\mathbb{Z}[i] = \{a + bi \mid a, b \in \mathbb{Z}\} \subseteq \mathbb{C}$ and $\mathbb{Z}[\sqrt{-5}] = \{a + b\sqrt{-5} \mid a, b \in \mathbb{Z}\} \subseteq \mathbb{C}$.

Using rings and ideals we will prove Fermat's famous two-square theorem: a prime number $\equiv 1 \,(\mathrm{mod}\ 4)$ is the sum of two unique squares (e.g. $13 = 2^2 + 3^2$). We will also show how computing the two squares given the prime number is related to quadratic residues and the Euclidean algorithm.

The first part of this chapter is a little on the heavy side concerning new concepts and definitions. Do not despair. None of them is really difficult. Make sure you study all examples intensively and link them to the concepts. Applications of the theory begin in subsection 3.5.5.

3.1 Definition

A *ring* is an abelian group $(R, +)$ with an additional composition $\cdot : R \times R \to R$ called multiplication. Multiplication satisfies the following for every $x, y, z \in R$:

(i) $(x \cdot y) \cdot z = x \cdot (y \cdot z)$;
(ii) there exists an element $1 \in R$ such that $1 \cdot x = x \cdot 1 = x$;
(iii) $x \cdot (y + z) = x \cdot y + x \cdot z$ and $(y + z) \cdot x = y \cdot x + z \cdot x$.

We will usually leave out $\cdot$ in $x \cdot y$ and simply write xy. The neutral element in the abelian group $(R, +)$ is denoted 0.

Definition 3.1.1   Below we list some of the most important definitions concerning a ring R.

(i) A subset $S \subseteq R$ of a ring R is called a *subring* if S is a subgroup of $(R, +)$, $1 \in S$ and $xy \in S$ if $x, y \in S$.
(ii) An element $x \in R \setminus \{0\}$ is called a *zero divisor* if there exists $y \in R \setminus \{0\}$ such that $xy = 0$ or $yx = 0$.
(iii) An element $x \in R$ is called a *unit* if there exists $y \in R$ such that $xy = yx = 1$. In this case y is denoted x^{-1} and called the inverse of x. The set of units in R is denoted R^*.
(iv) R is called *commutative* if $xy = yx$ for every $x, y \in R$.

The multiplication in R makes R^* into a group. If $R \neq \{0\}$ then $0 \notin R^*$. The group of units in a commutative ring R is an abelian group.

Example 3.1.2   The integers $\mathbb{Z}$ with addition and multiplication form in some sense the most natural commutative ring (later we will see that there is a unique ring homomorphism from $\mathbb{Z}$ into any commutative ring). Notice that $\mathbb{Z}^* = \{-1, 1\}$. An example of a non-commutative ring is provided by the 2×2 matrices

$$\mathrm{Mat}_2(\mathbb{R}) = \left\{ \begin{pmatrix} a & b \\ c & d \end{pmatrix} \,\middle|\, a, b, c, d \in \mathbb{R} \right\}$$

with real entries. Here the addition is the usual addition of matrices

$$\begin{pmatrix} a & b \\ c & d \end{pmatrix} + \begin{pmatrix} a' & b' \\ c' & d' \end{pmatrix} = \begin{pmatrix} a + a' & b + b' \\ c + c' & d + d' \end{pmatrix},$$

and the multiplication is the usual multiplication of matrices,

$$\begin{pmatrix} a & b \\ c & d \end{pmatrix} \begin{pmatrix} a' & b' \\ c' & d' \end{pmatrix} = \begin{pmatrix} aa' + bc' & ab' + bd' \\ ca' + dc' & cb' + dd' \end{pmatrix}.$$

The complex numbers $\mathbb{C}$ form a ring containing the integers $\mathbb{Z}$ as a subring. Again $\mathbb{C}$ is a subring of the ring of quaternions $\mathbb{H} = \{a + bi + cj + dk \mid a, b, c, d \in \mathbb{R}\}$, where addition is given component-wise and multiplication can be computed by the relations $i^2 = j^2 = k^2 = ijk = -1$. Using these relations one can deduce the composition table

·	1	i	j	k
1	1	i	j	k
i	i	-1	k	$-j$
j	j	$-k$	-1	i
k	k	j	$-i$	-1

for the multiplication. So $\mathbb{H}$ is a non-commutative ring with a highly intricate multiplication. Its discoverer, William R. Hamilton (1805–65) wrote

> Tomorrow will be the fifteenth birthday of the Quaternions. They started into life, or light, full grown, on the 16th of October, 1843, as I was walking with Lady Hamilton in Dublin, and came up to Brougham Bridge. That is to say, I then and there felt the galvanic circuit of thought closed, and the sparks which fell from it were the fundamental equations between I, J and K; *exactly such* as I have used them ever since. I pulled out, on the spot, a pocketbook, which still exists, and made an entry, on which, *at the very moment*, I felt that it might be worth my while to expend the labour of at least ten (or it might be fifteen) years to come. But then it is fair to say that this was because I felt a problem to have been at that moment solved, an intellectual *want relieved*, which had *haunted* me for at least *fifteen years* before.

Even though non-commutative rings are extremely interesting, we shall limit ourselves to commutative rings in the rest of this book. So, from this point onwards a ring will always refer to a commutative ring.

The rational and complex numbers are both examples of rings R satisfying $R^* = R \setminus \{0\}$. A ring R with $R^* = R \setminus \{0\}$ is called a *field*. If $K \subseteq L$ are fields and K is a subring of L then K is called a *subfield* of L and L is called an *extension field* of K. A *domain* is a ring $R \neq \{0\}$ with no zero divisors. Let us record the first basic properties about domains and fields.

Proposition 3.1.3 *Let R be a domain and $a, x, y \in R$. If $a \neq 0$ and $ax = ay$ then $x = y$.*

Proof. If $ax = ay$ then $a(x - y) = 0$. Since $a \neq 0$, this means that $x - y = 0$; thus $x = y$.   □

Proposition 3.1.4   *Let F be a field. Then F is a domain.*

Proof.   Suppose that $x, y \in F$, $x \neq 0$ and $xy = 0$. We must prove that $y = 0$. Since $x \neq 0$, there exists $x^{-1} \in F$ such that $x^{-1}x = 1$. This means that $0 = x^{-1}0 = x^{-1}(xy) = (x^{-1}x)y = y$.   □

The set of integers $\mathbb{Z}$ is a subring of the rational numbers $\mathbb{Q}$ with the usual addition and multiplication and $\mathbb{Z}^* = \{1, -1\}$. So $\mathbb{Z}$ is a domain that is not a field. The ring of rational numbers $\mathbb{Q}$ is a field, since every fraction $a/b \neq 0$ can be inverted: $(a/b)\,(b/a) = 1$. The ring of rational numbers $\mathbb{Q}$ is a subfield of the real numbers $\mathbb{R}$ and the real numbers form a subfield of the complex numbers $\mathbb{C}$ that is an extension field of $\mathbb{R}$.

Example 3.1.5   Consider the subset

$$\mathbb{Q}(i) = \{a + bi \mid a, b \in \mathbb{Q}\}$$

of $\mathbb{C}$. The usual rules, $(a + bi) + (c + di) = (a + c) + (b + d)i$ and $(a + bi)(c + di) = (ac - bd) + (ad + bc)i$, for adding and multiplying complex numbers imply that $\mathbb{Q}(i)$ is a subring of $\mathbb{C}$. If $z = a + bi$ is a non-zero element of $\mathbb{Q}(i)$ then

$$\frac{1}{z} = \frac{a - bi}{(a + bi)(a - bi)} = \frac{a}{a^2 + b^2} - \frac{b}{a^2 + b^2}i$$

and it follows that $\mathbb{Q}(i)$ is a field. It is an extension field of $\mathbb{Q}$ and a subfield of $\mathbb{C}$. Recall that $|z|^2 = z\bar{z}$, where $|z|$ is the modulus and $\bar{z}$ the complex conjugate of $z \in \mathbb{C}$. We call $|z|^2$ the norm of $z \in \mathbb{C}$ and denote it $N(z)$. Notice that

$$N(z_1 z_2) = N(z_1)N(z_2) \tag{3.1}$$

for $z_1, z_2 \in \mathbb{C}$. If $z = a + bi$ then $N(z) = (a + bi)(a - bi) = a^2 + b^2$. Inside $\mathbb{Q}(i)$ we have the subring $\mathbb{Z}[i] = \{a + bi \mid a, b \in \mathbb{Z}\}$, which is called the ring of *Gaussian integers*. Note that $N(z) \in \mathbb{N}$ if $z \in \mathbb{Z}[i]$. The property (3.1) of N implies that an element $z \in \mathbb{Z}[i]$ is a unit if and only if $N(z) = 1$: if z is a unit then there exists $y \in \mathbb{Z}[i]$ such that $zy = 1$, and (3.1) gives $1 = N(zy) = N(z)N(y)$, so that $N(z) = 1$. However, if $z = a + bi$ and $N(z) = (a + bi)(a - bi) = a^2 + b^2 = 1$ then $zy = 1$, where $y = a - bi \in \mathbb{Z}[i]$ and z is a unit. Using this characterization of the units one finds that $\mathbb{Z}[i]^* = \{1, -1, i, -i\}$.

Prime numbers in $\mathbb{Z}$ are not necessarily "prime numbers" in $\mathbb{Z}[i]$: for 5, for example, we have the factorization $5 = (1 + 2i)(1 - 2i)$ in $\mathbb{Z}[i]$. We will have a good deal more to say about this phenomenon later in the chapter.

3.1.1 Ideals

An *ideal* in a ring R is a subgroup I of $(R, +)$ such that $\lambda x \in I$ for every $\lambda \in R$ and $x \in I$. Notice that R itself is an ideal and that a given ideal I is the whole ring R if and only if $1 \in I$ (see Exercise 3.4).

Let $r_1, \ldots, r_n \in R$. Then the subset

$$\langle r_1, \ldots, r_n \rangle = \{\lambda_1 r_1 + \cdots + \lambda_n r_n \mid \lambda_1, \ldots, \lambda_n \in R\}$$

is an ideal in R (see Exercise 3.5). If I is an ideal in R and there exist $r_1, \ldots, r_n \in R$ such that $I = \langle r_1, \ldots, r_n \rangle$, we say that I is (finitely) *generated* by $r_1, \ldots, r_n \in R$. Notice that $\langle r_1, \ldots, r_n \rangle \subseteq I$ if $r_1, \ldots, r_n \in I$ (see Exercise 3.6).

Remark 3.1.6 It also makes sense to talk about an ideal generated by infinitely many elements. One defines this as follows. Let M be any subset of R. Then the ideal generated by M is

$$\langle f \mid f \in M \rangle = \{a_1 f_1 + \cdots + a_n f_n \mid n \in \mathbb{N},\ a_1, \ldots, a_n \in R,\ f_1, \ldots, f_n \in M\}.$$

Remark 3.1.7 Let I and J be ideals in a ring R.

(i) Then $I \cap J$ and $I + J = \{i + j \mid i \in I,\ j \in J\}$ are also ideals in R.
(ii) The product IJ of I and J is defined to be the ideal generated by $\{ij \mid i \in I,\ j \in J\}$ according to Remark 3.1.6. This has an obvious generalization to a finite number of ideals. Notice that $IJ \subseteq I \cap J$.

Remark 3.1.8 An ideal in a field F is either $\langle 0 \rangle$ or F itself. If $I \neq \langle 0 \rangle$ is an ideal in F and $a \in I \setminus \{0\}$ we can find $b \in F$ such that $ba = 1$. By the definition of an ideal, $1 = ba \in I$. This implies that $I = F$.

An ideal I in R that can be generated by one element is called a *principal ideal*. In this case there exists $d \in R$ such that $I = \langle d \rangle$.

Definition 3.1.9 A domain in which every ideal is a principal ideal is called a *principal ideal domain*.

Proposition 3.1.10 *The ring $\mathbb{Z}$ is a principal ideal domain.*

Proof. A subgroup of $\mathbb{Z}$ can be written $d\mathbb{Z}$ for $d \in \mathbb{Z}$ (see Proposition 2.2.3). This shows that every subgroup is a principal ideal. Since an ideal is in particular a subgroup this finishes the proof. □

Let us study ideals in the ring of Gaussian integers. We have already seen that the norm function $N(a + bi) = a^2 + b^2$ plays a central role. In the following crucial result we use a special property of the norm function, which will be formalized later in the notion of a Euclidean domain.

Theorem 3.1.11 *The ring of Gaussian integers $\mathbb{Z}[i]$ is a principal ideal domain.*

Proof. Let I be a non-zero ideal in $\mathbb{Z}[i]$. Choose among the non-zero elements in I an element $d = a + bi \in I$ such that $N(d) = a^2 + b^2$ is minimal. Now suppose that $z \in I$; then, computing in $\mathbb{C}$ we get $z/d = q_1 + q_2 i$, where $q_1, q_2 \in \mathbb{Q}$. A point in the complex plane is at most $\sqrt{2}/2$ away from a point with integer real and imaginary parts (why?). Therefore we may choose an element $q = c + di \in \mathbb{Z}[i]$ such that $|z/d - q|^2 < 1$ or, using the norm given in Example 3.1.5,

$$N(z/d - q) < 1. \tag{3.2}$$

Multiplying both sides of (3.2) by $N(d)$ we get $N(z - qd) < N(d)$, using (3.1). Since $z - qd \in I$, we must have that $z = qd$ by the construction of d. Thus $I \subseteq \langle d \rangle$. The other inclusion holds since $d \in I$, so we have proved that I is a principal ideal. □

However, the ring $\mathbb{Z}[\sqrt{-5}] = \{x + y\sqrt{-5} \mid x, y \in \mathbb{Z}\}$ contains ideals that are not principal. This will be revealed later in this chapter.

3.2 Quotient rings

Let I be an ideal in a ring R. Then I is in particular a subgroup of the abelian group $(R, +)$, and the set $R/I = \{[x] \mid x \in R\}$ of left cosets $[x] = x + I$ of I with respect to $+$ is an abelian group (recall that $[x] = [y]$ if and only if $x - y \in I$). We can make R/I into a ring in a very natural way by defining addition and multiplication as follows:

(i) $[x] + [y] = [x + y]$ for every $[x], [y] \in R/I$,
(ii) $[x][y] = [xy]$ for every $[x], [y] \in R/I$.

It is built into the definition of an ideal that these operations are independent of the choice of the element in the left coset. Suppose that $[x] = [x']$ and $[y] = [y']$. For the definition to be independent of the choice of element, we need that $[x + y] = [x' + y']$ and $[xy] = [x'y']$. We already know that $[x + y] = [x' + y']$, since composition in the quotient group $(R/I, +)$ is well defined. As $xy - x'y' = x(y - y') + y'(x - x') \in I$, it follows that $xy - x'y' \in I$ and therefore that $[xy] = [x'y']$. Notice how all this is inspired by Proposition 1.3.4. The new ring R/I is called the quotient ring of R by I and has $[0]$ and $[1]$ playing the role of 0 and 1. Notice that $[x] = 0$ in R/I if and only if $x \in I$.

3.2.1 Quotient rings of $\mathbb{Z}$

An ideal in $\mathbb{Z}$ is a principal ideal $\langle d \rangle$ generated by a natural number d. Two elements $x, y \in \mathbb{Z}$ represent the same element $[x] = [y]$ in $\mathbb{Z}/d\mathbb{Z}$ if and only if $x - y \in d\mathbb{Z}$ if and only if $d \mid x - y$. One way of thinking of the elements in $\mathbb{Z}/d\mathbb{Z}$ is as represented by the remainders by division with d, $\{[0], [1], [2], [3], \ldots, [d-1]\}$.

Example 3.2.1 If $d = 6$ then $\mathbb{Z}/6\mathbb{Z} = \{[0], [1], [2], [3], [4], [5]\}$. Here we have $[4] + [4] = [2]$ and $[3][4] = [0]$.

Proposition 3.2.2 *Suppose that d is a positive integer. Then the group of units $(\mathbb{Z}/d\mathbb{Z})^*$ is an abelian group with $\varphi(d)$ elements.*

Proof. Let us check that a coset $[x] = x + d\mathbb{Z}$ is a unit if and only if $\gcd(x, d) = 1$. If $\gcd(x, d) = 1$ then we can find $\lambda, \mu \in \mathbb{Z}$ such that $\lambda x + \mu d = 1$. Therefore $[\lambda x + \mu d] = [\lambda x] + [\mu d] = [\lambda][x] = [1]$, so that x is a unit. However, if $[x]$ is a unit in $\mathbb{Z}/d\mathbb{Z}$ then there exists an element $[\lambda] \in \mathbb{Z}/d\mathbb{Z}$ such that $[\lambda][x] = [\lambda x] = [1]$. Thus $\lambda x - 1 \in d\mathbb{Z}$ and we can find $\mu \in \mathbb{Z}$ such that $\lambda x - 1 = \mu d$. This implies that $\gcd(x, d) = 1$. $\qquad\square$

Notice the connection with subsection 2.3.2, where we constructed $(\mathbb{Z}/d\mathbb{Z})^*$ without using the ring structure of $\mathbb{Z}/d\mathbb{Z}$.

Proposition 3.2.3 *Let $n \in \mathbb{N}$. Then $\mathbb{Z}/n\mathbb{Z}$ is a field if and only if n is a prime number. If n is a composite number then $\mathbb{Z}/n\mathbb{Z}$ is not a domain.*

Proof. Assume that $n > 0$. By Proposition 3.2.2 we have $|(\mathbb{Z}/n\mathbb{Z})^*| = \varphi(n)$. Since $|\mathbb{Z}/n\mathbb{Z}| = n$, this shows that $\mathbb{Z}/n\mathbb{Z}$ is a field if and only if $\varphi(n) = n - 1$.

This last condition holds if and only if n is a prime number. If n is a composite number, we may write $n = ab$, where $1 < a, b < n$. This means that $[a] \neq [0]$ and $[b] \neq [0]$ in $\mathbb{Z}/n\mathbb{Z}$, but $[a][b] = [n] = [0]$, so that $\mathbb{Z}/n\mathbb{Z}$ is not a domain.   □

Remark 3.2.4   What happens if $n = 0$ in Proposition 3.2.3?

Definition 3.2.5   The field $\mathbb{Z}/p\mathbb{Z}$ is denoted $\mathbb{F}_p$, where p is a prime number.

3.2.2 Prime ideals

Suppose that I is an ideal in a ring R. When is the quotient ring R/I a domain? When is R/I a field? Suppose that R/I is a domain. Then $R/I \neq 0$ and $[x][y] = 0$ implies $[x] = 0$ or $[y] = 0$ for every $[x], [y] \in R/I$. In terms of the ideal I this means that

$$I \neq R \text{ and } xy \in I \quad \text{implies} \quad x \in I \text{ or } y \in I$$

for every $x, y \in R$. An ideal satisfying this condition is called a *prime ideal*. Conversely, if $I \subseteq R$ is a prime ideal then R/I is a domain (see Exercise 3.21). Thus we end up with the following proposition.

Proposition 3.2.6   *An ideal $I \subseteq R$ is a prime ideal if and only if R/I is a domain.*

3.2.3 Maximal ideals

Suppose that R/I is a field. This means that $R/I \neq 0$ and that for every non-zero element $[x] \in R/I$ there exists $[y] \in R/I$ such that $[x][y] = [xy] = [1]$.

In terms of the ideal I, this means that for every $x \notin I$ there exists $y \in R$ such that $xy - 1 \in I$. Suppose that J is another ideal such that $I \subseteq J \subseteq R$. If $x \in J \setminus I$ then we may find $y \notin I$ such that $xy - 1 \in I \subseteq J$. But since xy is in J (as $x \in J$) it follows that $1 = -(xy - 1) + xy \in J$. This means that $J = R$. We have proved that if R/I is a field then I is an ideal satisfying the following:

$$\text{if } I \subsetneq J \text{ then } J = R,$$

where J is an ideal of R. An ideal satisfying this condition is called a *maximal ideal* (maximal among the ideals properly contained in R).

If $I \subseteq R$ is a maximal ideal then R/I is a field. This can be seen as follows. If $[x] \in R/I$ is a non-zero element then $x \notin I$. The subset $I + Rx = \{i + rx \mid i \in I, r \in R\}$ is an ideal in R. Since $I \subsetneq I + Rx$, we must have that $I + Rx = R$. Therefore $1 \in I + Rx$. So we may write $1 = m + rx$ for suitable $m \in I, r \in R$. Going to R/I we get $[1] = [r][x]$, so that $[x]$ is a unit in R/I. We end up with the following proposition.

Proposition 3.2.7 *An ideal $I \subseteq R$ is a maximal ideal if and only if R/I is a field.*

Remark 3.2.8 A maximal ideal is a prime ideal, because a field is a domain (Proposition 3.1.4).

Example 3.2.9 The ring $\mathbb{Z}$ is a principal ideal domain. This means that every ideal in $\mathbb{Z}$ has the form $\langle d \rangle = d\mathbb{Z}$ for some $d \in \mathbb{Z}$. Which of these are maximal? Suppose that p is a prime number and that $\langle p \rangle = p\mathbb{Z}$ is contained in another ideal $\langle d \rangle = d\mathbb{Z}$ in $\mathbb{Z}$. Then $p \in \langle d \rangle$. Therefore d divides p and so $d = \pm 1$ or $d = \pm p$. This implies that $\langle d \rangle = \mathbb{Z}$ or $\langle d \rangle = p\mathbb{Z}$, proving that $\langle p \rangle$ is a maximal ideal. The ideal $\langle 0 \rangle$ is a prime ideal that is not a maximal ideal. An ideal $\langle m \rangle$ generated by a composite number $m = ab$, where $a, b \neq \pm 1$ is not a prime ideal, since $ab \in \langle m \rangle$ but $a \notin \langle m \rangle$ and $b \notin \langle m \rangle$.

3.3 Ring homomorphisms

A map $f : R \to S$ between two rings R and S is called a *ring homomorphism* if it is a group homomorphism from $(R, +)$ to $(S, +)$, $f(xy) = f(x)f(y)$ for every $x, y \in R$ and $f(1) = 1$. A bijective ring homomorphism is called a *ring isomorphism*. If R and S are rings and there exists a ring isomorphism $f : R \to S$, we say that R and S are isomorphic. This is denoted $R \cong S$.

Example 3.3.1 The map $R \to R/I$ given by $r \mapsto [r]$ is a (surjective) ring homomorphism. This follows from the way we defined addition and multiplication in R/I.

The kernel Ker $f = \{r \in R \mid f(r) = 0\} \subseteq R$ of f (as a group homomorphism) is an ideal of R and the image $f(R)$ is a subring of S (see Exercise 3.11). The isomorphism theorem for rings follows almost immediately from the analogue for groups (see Theorem 2.5.1).

Proposition 3.3.2   *Let R and S be rings and $f : R \to S$ a ring homomorphism with kernel $K = \mathrm{Ker}(f)$. Then*

$$\tilde{f} : R/K \to f(R)$$

given by $\tilde{f}(r + K) = f(r)$ is a well defined map and a ring isomorphism.

Proof.   We already know that $\tilde{f}$ is a well defined map and an isomorphism of abelian groups by Theorem 2.5.1. It remains to check that it is a ring homomorphism. Clearly $\tilde{f}(1 + K) = f(1) = 1$. Since

$$\begin{aligned}
\tilde{f}((x + K)(y + K)) &= \tilde{f}(xy + K) \\
&= f(xy) = f(x)f(y) \\
&= \tilde{f}(x + K)\tilde{f}(y + K)
\end{aligned}$$

for $x, y \in R$, it follows that $\tilde{f}$ is a ring homomorphism. $\square$

3.3.1   The unique ring homomorphism from $\mathbb{Z}$

Lemma 3.3.3   *For every ring R, there is a unique ring homomorphism $f : \mathbb{Z} \to R$.*

Proof.   A ring homomorphism $f : \mathbb{Z} \to R$ is in particular a group homomorphism $f : (\mathbb{Z}, +) \to (R, +)$ with $f(1) = 1$. This last condition says that $f = f_1$ in the notation of Section 2.6. So f is unique. We just need to show that $f = f_1 : \mathbb{Z} \to R$ is a ring homomorphism. In other words we must show that $f(mn) = f(m)f(n)$ for $m, n \in \mathbb{Z}$. We can assume that $m, n > 0$, since $f(-m) = f((-1)m) = f(-1)f(m) = -f(m)$. Now the result follows if $x(y + z) = xy + xz$ $(x, y, z \in R)$ is applied successively: a sum of m copies of 1 multiplied by a sum of n copies of 1 is a sum of mn copies of 1 (here $1 \in R$). $\square$

Remark 3.3.4   Let $f : \mathbb{Z} \to R$ denote the unique ring homomorphism for a given ring R. For $n \geq 0$, one thinks of $f(n)$ as

$$f(n) = 1 + 1 + \cdots + 1,$$

a sum of n copies of $1 \in R$. Given the unique ring homomorphism $f : \mathbb{Z} \to R$ *it makes sense to view integers as elements in any ring. When $n \in \mathbb{Z}$ and we write $n \in R$ we are referring to the element $f(n)$ of R.*

Let R be a ring. Let $\mathrm{ord}\,(1)$ denote the order of 1 in $(R, +)$. This turns out to be a fundamental invariant of R. If $\mathrm{ord}\,(1)$ is infinite then R is said to have *characteristic* zero. If $\mathrm{ord}\,(1)$ is finite R is said to have finite characteristic $\mathrm{ord}\,(1)$. So the characteristic of R is n_1, where $n_1 \in \mathbb{N}$ and $n_1\mathbb{Z} = \mathrm{Ker}\,f_1$ in the notation of Section 2.6. The characteristic of R is denoted char R. In the positive-characteristic case one usually thinks of char R as the smallest natural number n for which $1 + \cdots + 1$ (n times) $= 0$ in R.

The ring $\mathbb{Z}$ of integers has characteristic zero. The same is true for $\mathbb{Q}$, $\mathbb{R}$. But char $\mathbb{Z}/n\mathbb{Z} = n$ for $n \in \mathbb{N}$.

Lemma 3.3.5   *Let R be a ring. Then there is an injective ring homomorphism*

$$\mathbb{Z}/n\mathbb{Z} \to R,$$

where $n = $ char R.

Proof.   Let $f : \mathbb{Z} \to R$ be the unique ring homomorphism. Then $f(\mathbb{Z}) = S$ is a subring of R and $\mathrm{Ker}\,(f) = n\mathbb{Z}$ for $n = $ char R. The isomorphism theorem for rings (Proposition 3.3.2) says that we have a ring isomorphism

$$\mathbb{Z}/n\mathbb{Z} \to S.$$

But this means that we have the desired injective ring homomorphism $\mathbb{Z}/n\mathbb{Z} \to S \subseteq R$. $\qquad\square$

Remark 3.3.6   In the situation of Lemma 3.3.5 we say that $\mathbb{Z}/n\mathbb{Z}$ is contained in R, since it is isomorphic to a subring in R.

Proposition 3.3.7   *Let R be a domain. Then* char R *is either zero or a prime number. If R is finite then R is a field and* char R *is a prime number.*

Proof.   Let $n = $ char R. We know that $\mathbb{Z}/n\mathbb{Z}$ is a subring of R by Lemma 3.3.5. This means in particular that $\mathbb{Z}/n\mathbb{Z}$ is a domain, being a subring of a domain. In this way n must be zero or a prime number by Proposition 3.2.3. If R is a finite domain then $n > 0$ (if $n = 0$, R would contain $\mathbb{Z}$ as a subring) and n must be a prime number. A finite domain is a field (see Exercise 3.23). $\qquad\square$

3.3.2 Freshman's Dream

The title of this subsection refers to certain beginners' mistakes in calculus exercises: for example, that the sine of a sum of two angles $\sin(x + y)$ is equal to $\sin(x) + \sin(y)$ or that $(x + y)^5$ is equal to $x^5 + y^5$ for $x, y \in \mathbb{R}$. Of course, one has to insert intermediate terms coming from the binomial formula to evaluate $(x + y)^5$. Let us state a general version of the binomial formula.

Lemma 3.3.8 *Let R be a ring and a, b two elements in R. Then*

$$(a + b)^n = a^n + \binom{n}{1}a^{n-1}b + \cdots + \binom{n}{n-1}ab^{n-1} + b^n$$

for $n \in \mathbb{N}$.

Proof. This can be proved using induction. The case $n = 1$ is clear. Assume that

$$(a + b)^n = a^n + \binom{n}{1}a^{n-1}b + \cdots + \binom{n}{n-1}ab^{n-1} + b^n;$$

then $(a + b)^{n+1} = (a + b)^n(a + b)$. Using $ab = ba$ and

$$\binom{n}{i} + \binom{n}{i-1} = \binom{n+1}{i}$$

for $i = 1, \ldots, n$, the result follows. $\square$

Notice that the binomial coefficients in Lemma 3.3.8 are considered as elements in the ring R through the unique ring homomorphism $\mathbb{Z} \to R$. We will keep using this convention. Now for the main insight, which looks innocent but is incredibly powerful.

Theorem 3.3.9 (Freshman's Dream) *Let R be a ring of prime characteristic p. Then*

$$(x + y)^{p^r} = x^{p^r} + y^{p^r}$$

for every $x, y \in R$ and $r \in \mathbb{N}$.

Proof. Since char $R = p$, the kernel of the unique ring homomorphism $\mathbb{Z} \to R$ is $p\mathbb{Z}$. As p divides the binomial coefficients

$$\binom{p}{1}, \quad \binom{p}{2}, \quad \ldots \quad \binom{p}{p-1},$$

by Exercise 1.30(i), it follows that $\binom{p}{i} = 0$ in R when $i = 1, \ldots, p - 1$. Using Lemma 3.3.8, this shows that $(x + y)^p = x^p + y^p$ in R. Now conclude by induction for $r > 1$ that

$$(x + y)^{p^r} = ((x + y)^p)^{p^{r-1}} = (x^p + y^p)^{p^{r-1}} = (x^p)^{p^{r-1}} + (y^p)^{p^{r-1}} = x^{p^r} + y^{p^r}.$$

                                                                                   $\square$

Freshman's Dream is one of the most useful facts in algebra. Doing mathematics in a universe where this kind of linearity is possible is a dream come true. Already in the following chapter on polynomials, Freshman's Dream will become an indispensable tool especially in proving the law of quadratic reciprocity.

Remark 3.3.10 Notice that if R is a ring of prime characteristic p then Theorem 3.3.9 shows that the map $F : R \rightarrow R$ given by $F(x) = x^p$ is a ring homomorphism. It is called the Frobenius map after G. Frobenius (1849–1917).

3.4 Fields of fractions

If R is a domain then there is a very natural field Q and an injective ring homomorphism $R \rightarrow Q$. In a precise sense one may say that Q is the "smallest" field containing R. The field Q consists of fractions with a numerator in R and a denominator in $R \setminus \{0\}$. The situation is practically identical with the situation $R = \mathbb{Z}$ and $Q = \mathbb{Q}$ and the construction the same as in Appendix A.2.2. We let $M = R \times (R \setminus \{0\})$ and define $Q = M/\sim$, where $(a, s) \sim (b, t)$ if and only if $at = bs$. As in Appendix A.2.2 we let $\dfrac{a}{s}$ denote the equivalence class containing $(a, s) \in M$. Then

$$\frac{a}{s}\frac{b}{t} = \frac{ab}{st}, \qquad \frac{a}{s} + \frac{b}{t} = \frac{at + bs}{st},$$

are well defined operations and they make Q into a ring, where

$$0 = \frac{0}{a} \qquad \text{and} \qquad 1 = \frac{a}{a}$$

for every $a \in R \setminus \{0\}$. Notice that Q is a field, since

$$\frac{a}{s} \neq 0$$

in Q means that $a \neq 0$. In this case

$$\frac{s}{a} \in Q \qquad \text{and} \qquad \frac{a}{s}\frac{s}{a} = \frac{as}{as} = 1$$

in Q. Furthermore, Q comes with an injective ring homomorphism $i : R \rightarrow Q$ given by

$$i(a) = \frac{a}{1}.$$

The field Q is called the *field of fractions* of R. The following proposition states formally that it is the "smallest" field containing R.

Proposition 3.4.1   *Let R be a domain with field of fractions Q, let L be a field and let $\varphi : R \to L$ be an injective ring homomorphism. Then there exists a unique injective ring homomorphism $\bar{\varphi} : Q \to L$ such that $\bar{\varphi} \circ i = \varphi$.*

Proof.   If $\bar{\varphi} \circ i = \varphi$ then we must have

$$1 = \bar{\varphi}\left(\frac{s}{1}\frac{1}{s}\right) = \bar{\varphi}\left(\frac{s}{1}\right)\bar{\varphi}\left(\frac{1}{s}\right) = \varphi(s)\bar{\varphi}\left(\frac{1}{s}\right),$$

where $s \in R \setminus \{0\}$. So there is only one way of defining $\bar{\varphi}$, provided that $\bar{\varphi} \circ i = \varphi$:

$$\bar{\varphi}\left(\frac{a}{s}\right) = \bar{\varphi}\left(\frac{a}{1}\right)\bar{\varphi}\left(\frac{1}{s}\right) = \varphi(a)\varphi(s)^{-1}.$$

This is well defined: if

$$\frac{a}{s} = \frac{b}{t}$$

then $at = bs$. Therefore $\varphi(a)\varphi(t) = \varphi(b)\varphi(s)$ and $\varphi(a)\varphi(s)^{-1} = \varphi(b)\varphi(t)^{-1}$. Let us prove that $\bar{\varphi}$ really is a ring homomorphism. Proving that $\bar{\varphi}$ preserves multiplication is left to the reader. Below we prove that $\bar{\varphi}$ preserves addition:

$$\bar{\varphi}\left(\frac{a}{s} + \frac{b}{t}\right) = \bar{\varphi}\left(\frac{at + bs}{st}\right)$$
$$= (\varphi(a)\varphi(t) + \varphi(b)\varphi(s))\varphi(s)^{-1}\varphi(t)^{-1}$$
$$= \varphi(a)\varphi(s)^{-1} + \varphi(b)\varphi(t)^{-1}$$
$$= \bar{\varphi}\left(\frac{a}{s}\right) + \bar{\varphi}\left(\frac{b}{t}\right).$$

To prove that $\bar{\varphi}$ is injective it is enough to show that $\mathrm{Ker}\,(\bar{\varphi}) = \{0\}$. Suppose that

$$\bar{\varphi}\left(\frac{a}{s}\right) = \varphi(a)\varphi(s)^{-1} = 0.$$

Then $\varphi(a) = 0$ and therefore $a = 0$ since φ is injective. This proves that

$$\frac{a}{s} = 0$$

and therefore that $\mathrm{Ker}\,(\bar{\varphi}) = \{0\}$. $\square$

Corollary 3.4.2 *Let R be a domain contained in the field L. The smallest subfield in L containing R is*

$$K = \{as^{-1} \mid a \in R, s \in R \setminus \{0\}\}.$$

The field of fractions of R is isomorphic to K.

Proof. Let $a, b \in R$ and $s, t \in R \setminus \{0\}$. Then $(as^{-1})(bt^{-1}) = (ab)(ts)^{-1}$, $as^{-1} + bt^{-1} = (at + bs)(ts)^{-1}$ and $(as^{-1})^{-1} = sa^{-1}$ if $a \neq 0$. These formulas imply that K is a subfield of L. Any subfield of L containing R must contain as^{-1}, where $a \in R$ and $s \in R \setminus \{0\}$. So K is the smallest subfield containing R. Let Q be the field of fractions of R. Then the unique injective ring homomorphism $\bar{\varphi} : Q \to K$ of Proposition 3.4.1 is surjective, since it is given by

$$\bar{\varphi}\left(\frac{a}{s}\right) = as^{-1}$$

(notice that φ is the inclusion of R into L). It is therefore an isomorphism and Q becomes isomorphic to K. □

Example 3.4.3 The Gaussian integers $\mathbb{Z}[i]$ form a domain whose field of fractions is isomorphic to $\mathbb{Q}(i)$. This follows from Corollary 3.4.2 and Example 3.1.5.

3.5 Unique factorization

What is the analogue of a prime number in a general commutative ring? Is there such a thing as unique factorization? Saying that a "general prime number" should be an element x that cannot be factored except for the factorization $x = ab$, where a or b is a unit, is not enough. The key property turns out to be the generalization of the fact that if a prime number divides a product of two numbers then it divides one of them (this is Lemma 1.8.3). A unique factorization domain is a domain like $\mathbb{Z}$, where every non-zero element has a unique factorization into prime elements. The main result in this section is that a principal ideal domain is a unique factorization domain (Theorem 3.5.7). The proof is not difficult once you recall how we proved unique factorization into prime numbers for $\mathbb{Z}$ in Theorem 1.8.5. The only difference is in Lemma 3.5.5, which in a sense is an abstract version of Lemma 1.8.1. In the following we will assume that R is a domain.

3.5.1   Divisibility and greatest common divisor

Suppose that x, $y \in R$. If $x = ry$ for some $r \in R$, we say that y is a *divisor* of x. This is denoted $y \mid x$. Notice that $y \mid x$ if and only if $\langle x \rangle \subseteq \langle y \rangle$. If $x = uy$, where $u \in R^*$, then $\langle x \rangle = \langle y \rangle$. However, if $\langle x \rangle = \langle y \rangle$ then $x = ry$ and $y = sx$ for some $r, s \in R$. Therefore $x = r(sx) = (rs)x$. Since R is a domain we conclude that $rs = 1$ by Proposition 3.1.3 (if $x \neq 0$). This means that $r, s \in R^*$. Thus, $\langle x \rangle = \langle y \rangle$ implies that there exists $u \in R^*$ such that $x = uy$. In this case we say that x and y are *associated elements* of R.

An element $d \in R$ is a *greatest common divisor* of $a, b \in R$ if d is a common divisor of a and b and every common divisor of a and b divides d. Notice how this generalizes the greatest common divisor definition for the integers (see Section 1.4).

Let R be a principal ideal domain. For every $a, b \in R$ we know that there exists $d \in R$ such that $\langle a, b \rangle = \{xa + yb \mid x, y \in R\} = \langle d \rangle$. We claim that d is a greatest common divisor of a and b. Clearly d is a common divisor of a and b since $\langle a \rangle \subseteq \langle d \rangle$ and $\langle b \rangle \subseteq \langle d \rangle$. If e is a common divisor of a and b then $\langle e \rangle \supseteq \langle a, b \rangle = \langle d \rangle$. Thus e divides d and so d is a greatest common divisor of a and b.

3.5.2   Irreducible elements

An element $r \in R \setminus R^*$ is called *irreducible* if $r = ab$ for $a, b \in R$ implies that either a or b is a unit. Thus if r is an irreducible element and u is a unit then ur is also an irreducible element. A non-zero element $x \in R \setminus R^*$ is said to have a *factorization into irreducible elements* if there exist irreducible elements $p_1, \ldots, p_r \in R$ such that

$$x = p_1 \cdots p_r.$$

Now x is said to have *unique factorization into irreducible elements* if for any other irreducible factorization

$$x = q_1 \cdots q_s,$$

every p_i for $i = 1, \ldots, r$ divides q_j for some $j = 1, \ldots, s$ (this implies that $p_i = uq_j$, where u is a unit). In particular we have $r = s$ by Proposition 3.1.3. A domain R such that every non-zero element in $R \setminus R^*$ has unique factorization into irreducible elements is called *a unique factorization domain*.

Example 3.5.1   The irreducible elements in the ring of integers are $\pm p$, where p is a prime number. So $\mathbb{Z}$ is a unique factorization domain by Theorem 1.8.5.

We do not know yet whether the ring of Gaussian integers $\mathbb{Z}[i]$ is a unique factorization domain. This will follow once we have proved that a principal ideal domain is a unique factorization domain.

3.5.3 Prime elements

A non-zero element $p \in R \setminus R^*$ is called a *prime element* if $p \mid xy$ for $x, y \in R$ implies that $p \mid x$ or $p \mid y$.

Proposition 3.5.2 *A prime element is irreducible.*

Proof. Let p be a prime element. Suppose that $p = ab$. By definition of a prime element we can conclude that $p \mid a$ or $p \mid b$. Suppose that $p \mid a$. Then we can write $a = rp$ for some $r \in R$. This implies that $p = rpb$. Now Proposition 3.1.3 gives that b is a unit. Thus p is irreducible. □

Proposition 3.5.3 *Let R be a ring for which every non-zero element $x \in R \setminus R^*$ has a factorization into irreducible elements. Every irreducible element is a prime element in R if and only if R is a unique factorization domain.*

Proof. The "only if" part is identical to the proof of unique factorization for the integers (see Theorem 1.8.5). Suppose that $x \in R$ is a non-zero element with two factorizations:

$$x = p_1 \cdots p_r = q_1 \cdots q_s.$$

Now fix an irreducible element p_i from the left hand side. Since p_i is a prime element dividing a product $q_1 \cdots q_s$, it must divide some q_j (see Remark 1.8.4). Let us prove the "if" part. Assume that R is a unique factorization domain and let $p \in R$ be an irreducible element. Suppose that $p \mid ab$, where $a, b \in R$. We must prove that $p \mid a$ or $p \mid b$. Assume that $ab \neq 0$. Then a and b have factorizations into irreducible elements. Because of unique factorization, one of these factorizations must contain an irreducible element divisible by p. This proves the "if" part. □

Remark 3.5.4 The subset $\mathbb{Z}[\sqrt{-5}] = \{a + b\sqrt{-5} \mid a, b \in \mathbb{Z}\}$ is a subring of $\mathbb{C}$. In $\mathbb{Z}[\sqrt{-5}]$ we have two different factorizations of 6:

$$6 = 2 \cdot 3 = (1 + \sqrt{-5})(1 - \sqrt{-5}).$$

Both factorizations turn out to be irreducible, so that in this case the irreducible factorizations are not unique. Let us prove that 2 is an irreducible element of $\mathbb{Z}[\sqrt{-5}]$ that is not a prime element. From the above we see that $2 \mid (1 + \sqrt{-5})(1 - \sqrt{-5})$. But 2 does not divide either of these factors. Assume for example that $2 \mid 1 + \sqrt{-5}$. Then there exists $z \in \mathbb{Z}[\sqrt{-5}]$ such that $2z = 1 + \sqrt{-5}$. But this would show that

$$z = \frac{1}{2} + \frac{1}{2}\sqrt{-5} \notin \mathbb{Z}[\sqrt{-5}].$$

As in the case of the Gaussian integers (Example 3.1.5) the norm function $N(z) = z\bar{z}$ gives a function $N : \mathbb{Z}[\sqrt{-5}] \to \mathbb{N}$ such that $N(z_1 z_2) = N(z_1)N(z_2)$, where $z_1, z_2 \in \mathbb{Z}[\sqrt{-5}]$. If $z = x + y\sqrt{-5}$ then $N(x + y\sqrt{-5}) = (x + y\sqrt{-5})(x - y\sqrt{-5}) = x^2 + 5y^2$. Again it is easy to show that $z \in \mathbb{Z}[\sqrt{-5}]^*$ if and only if $N(z) = 1$. This gives that $z = x + y\sqrt{-5}$ is a unit if and only if $x = \pm 1$ and $y = 0$. To prove that 2 is an irreducible element, we assume that $2 = ab$, where $a = x + y\sqrt{-5}$ and $b = x' + y'\sqrt{-5}$. The crucial point is now to use the norm function. This gives $N(2) = 4 = N(a)N(b) = (x^2 + 5y^2)(x'^2 + 5y'^2)$, where $x, y, x', y' \in \mathbb{Z}$. We must have $y = y' = 0$ (why?), showing that one of a or b is a unit.

The following lemma is analogous to the statement that every non-zero integer is a product of prime numbers (Lemma 1.8.1).

Lemma 3.5.5 *Let R be a principal ideal domain and r a non-zero element. Then r has an irreducible factorization.*

Proof. An increasing sequence (a chain) of principal ideals $\langle a_1 \rangle \subseteq \langle a_2 \rangle \subseteq \langle a_3 \rangle \subseteq \cdots$ in R must stabilize: there is a step $N \in \mathbb{N}$ such that $\langle a_i \rangle = \langle a_{i+1} \rangle = \cdots$ for $i \geq N$. This is proved using that the union

$$\bigcup_{i=1}^{\infty} \langle a_i \rangle$$

is an ideal I in R (see Exercise 3.9, where you are also asked to show that the union of two ideals is not necessarily an ideal). From this we get $I = \langle d \rangle$, for some $d \in R$, since every ideal in R is principal. By definition of union, we must have $d \in \langle a_N \rangle$ for some N. Thus $\langle d \rangle \subseteq \langle a_N \rangle$ showing that $\langle a_i \rangle = \langle d \rangle$ for $i \geq N$.

Suppose that $r \in R \setminus R^*$ is a non-zero element that is not a product of irreducible elements. Then r is not irreducible. So we can write $r = r_1 s_1$ where $r_1, s_1 \notin R^*$. This means that $\langle r \rangle \subsetneq \langle r_1 \rangle$ and $\langle r \rangle \subsetneq \langle s_1 \rangle$. If both r_1 and s_1 are products of irreducible elements then so is r, contradicting our assumption. So, at least one of r_1 and s_1 is *not* a product of irreducible elements. Assume that r_1 is not a product of irreducible elements. Again we can write $r_1 = r_2 s_2$, where $r_2, s_2 \notin R^*$, $\langle r_1 \rangle \subsetneqq \langle r_2 \rangle$ and $\langle r_1 \rangle \subsetneqq \langle s_2 \rangle$. We may assume that r_2 is not a product of irreducible elements. Continuing in this way we obtain the infinite chain

$$\langle r \rangle \subsetneqq \langle r_1 \rangle \subsetneqq \langle r_2 \rangle \subsetneqq \cdots ,$$

where no r_i is a product of irreducible elements. This is a chain of ideals that does not stabilize, contradicting the first part of the proof. Thus every non-zero element r which is not a unit is a product of irreducible elements. $\square$

Proposition 3.5.6 *Suppose that R is a principal ideal domain that is not a field. An ideal $\langle x \rangle \subseteq R$ is a maximal ideal if and only if x is an irreducible element in R.*

Proof. If x is irreducible in R and $\langle x \rangle$ is contained in another ideal $\langle y \rangle$ then $x = ys$ for some $s \in R$. Since x is irreducible this implies that s or y is a unit. Thus $\langle y \rangle = \langle x \rangle$ or $\langle y \rangle = R$, showing that $\langle x \rangle$ is a maximal ideal. However, if $\langle x \rangle$ is a maximal ideal and $x = ys$ for some $y, s \in R$ then one of y and s must be a unit. If not, then $\langle x \rangle$ would be strictly contained in $\langle y \rangle$ since s is not a unit. Since y is not a unit, $\langle y \rangle$ must be strictly contained in R contradicting that $\langle x \rangle$ is a maximal ideal. $\square$

Theorem 3.5.7 *A principal ideal domain R is a unique factorization domain.*

Proof. In Lemma 3.5.5 we proved that every non-zero element has an irreducible factorization. The only thing missing is to prove that such a factorization is unique. This is accomplished, using Proposition 3.5.3, by proving that the irreducible elements are prime. Let $\pi \in R$ be an irreducible element such that $\pi \mid ab$ and $\pi \nmid a$. We will prove that $\pi \mid b$. That $\pi \nmid a$ implies $a \notin \langle \pi \rangle$ and therefore $\langle \pi, a \rangle \supsetneqq \langle \pi \rangle$. Since $\langle \pi \rangle$ is a maximal ideal, by Proposition 3.5.6, it follows that $\langle \pi, a \rangle = R = \langle 1 \rangle$. So we can find $x, y \in R$ such that $x\pi + ya = 1$. Now multiply both sides by b and get $xb\pi + yab = b$. Since $\pi \mid ab$, this shows that $\pi \mid b$. You should compare this to the proof of Corollary 1.5.10. They are practically identical except that here we have a more general framework. $\square$

Remark 3.5.8 The ring $\mathbb{Z}[\sqrt{-5}]$ is not a principal ideal domain since 2 is an irreducible element that is not a prime element (see Remark 3.5.4). In fact we can explicitly give an example of an ideal I in $\mathbb{Z}[\sqrt{5}]$ that is not a principal ideal. Let $I = \langle 2, 1 + \sqrt{-5} \rangle$. By explicit computation and a little rewriting one may prove that $I = \{(2a + b) + b\sqrt{-5} \mid a, b \in \mathbb{Z}\}$. This implies that $1 \notin I$, so that $I \neq R$. Let us assume that $I = \langle d \rangle$ for some $d \in \mathbb{Z}[\sqrt{-5}]$. Recall that $N(x + y\sqrt{-5}) = x^2 + 5y^2$, where $x, y \in \mathbb{Z}$. As d is not a unit, we get $N(d) > 1$. Since d divides every element of I, it must divide 2. Therefore $N(d) \mid N(2) = 4$ and we must have $N(d) = 4$, since $N(d) = 2$ is impossible. But $N(d) = 4$ means that we can assume that $d = 2$. But $1 + \sqrt{-5} \notin \langle 2 \rangle$, since $2 \nmid 1 + \sqrt{-5}$. We have proved that $\langle 2, 1 + \sqrt{-5} \rangle$ cannot be a principal ideal.

Suppose that we are given two elements a, b in a unique factorization domain. Suppose furthermore that we have found prime elements $p_1, \ldots, p_n$ such that

$$a = p_1^{r_1} \cdots p_n^{r_n},$$
$$b = p_1^{s_1} \cdots p_n^{s_n},$$

where $r_i, s_i \geq 0$. Then a greatest common divisor (see subsection 3.5.1) of a and b is given by

$$c = p_1^{t_1} \cdots p_n^{t_n},$$

where $t_i = \min(r_i, s_i)$; compare this with Remark 1.8.6. Usually it is very difficult (as for $\mathbb{Z}$) to compute prime factorizations of elements effectively. So relying on prime factorizations for finding a greatest common divisor may be a slow process. The Euclidean algorithm is in general much faster, but it does not necessarily exist in domains more general than $\mathbb{Z}$. If it does, there is a special term for the domain, as follows.

3.5.4 Euclidean domains

A domain R is called Euclidean if there exists a Euclidean function $N : R \setminus \{0\} \to \mathbb{N}$. A Euclidean function satisfies that for every $x \in R$, $d \in R \setminus \{0\}$, there exist $q, r \in R$ such that

$$x = qd + r,$$

where either $r = 0$ or $N(r) < N(d)$.

The ring of integers $\mathbb{Z}$ carries the absolute value $|\cdot| : \mathbb{Z} \to \mathbb{N}$ as a Euclidean function. Using Theorem 1.2.1 it is easy to verify that for every $x \in \mathbb{Z}$, $d \in \mathbb{Z} \setminus \{0\}$, there exist $q, r \in \mathbb{Z}$ such that

$$x = qd + r,$$

where $r = 0$ or $|r| < |d|$. After having seen the proof of Theorem 3.1.11, the following proposition should come as no surprise.

Proposition 3.5.9   *A Euclidean domain R is a principal ideal domain.*

Proof.   Let $I \subset R$ be a non-zero ideal in R and let $x \in I$ be a non-zero element such that $N(x)$ is minimal (compared with every $N(y)$, where $y \in I \setminus \{0\}$). We claim that $I = Rx$. Suppose that $y \in I$. Then we may find $q \in R$ such that

$$y = qx + r$$

where $r = 0$ or $N(r) < N(x)$. But as $r = y - qx \in I$, we must have $r = 0$ since $N(x)$ is minimal among $N(z)$, where z runs through the non-zero elements of I. This means that $y = qx$ and thus $I = Rx$.   □

Recall the definition in subsection 3.5.1 of a greatest common divisor along with the description of it in a principal ideal domain. A greatest common divisor of two elements in a Euclidean domain R can be found using the Euclidean algorithm (hence the term Euclidean). Here is how this works. Let $N : R \setminus \{0\} \to \mathbb{N}$ be a Euclidean function and suppose that $a, b \in R$. A greatest common divisor is a generator for the (principal) ideal $\langle a, b \rangle$.

If either a or b is zero, we are done. Suppose that both a and b are non-zero and that $N(a) \geq N(b)$. Then there exists $q \in R$ such that $a = qb + r$, where either $r = 0$ or $N(r) < N(b)$. We have $\langle a, b \rangle = \langle b, r \rangle$ since $r = a - qb \in \langle a, b \rangle$ and $a = qb + r \in \langle b, r \rangle$. Continue the procedure with $a = b$ and $b = r$ until one of a and b is zero. This will eventually happen, since we are strictly decreasing the maximum value of the norm function of a and b in each step. You should work out Exercise 3.29 to practice the Euclidean algorithm in the Gaussian integers with the norm function (you can do that before seeing the proof that $\mathbb{Z}[i]$ with the norm function is a Euclidean domain).

Remark 3.5.10   A principal ideal domain is not a Euclidean domain in general. The ring $R = \mathbb{Z}[\xi] = \{x + y\xi \mid x, y \in \mathbb{Z}\} \subseteq \mathbb{C}$, where $\xi = (1 + \sqrt{-19})/2$, is an example of a principal ideal domain that is not a Euclidean domain. One may prove that R cannot be a Euclidean domain using $R^* = \{\pm 1\}$. Proving that R is a principal ideal domain is more difficult.

We will see later that polynomial rings in one variable over a field are Euclidean domains (using the degree function). A nice fact is that the ring

of Gaussian integers is a Euclidean domain. This can actually be proved by
drawing circles in the complex plane.

3.5.5 Fermat's two-square theorem

A beautiful result due to Fermat says that a prime number $p \equiv 1 \,(\mathrm{mod}\ 4)$ is
the sum of two unique squares (e.g. $13 = 4 + 9$). We will prove this result
using unique factorization in the ring of Gaussian integers $\mathbb{Z}[i]$. Recall the
norm function $N : \mathbb{Z}[i] \to \mathbb{N}$ given by $N(a + bi) = (a + bi)(a - bi) = |a + bi|^2 = a^2 + b^2$. This function is an invaluable tool in reasoning about Gaussian
integers. Here is an example.

Proposition 3.5.11 *Let* $\pi = a + bi \in \mathbb{Z}[i]$ *be a Gaussian integer with*
$N(\pi) = p$, *where* p *is a prime number. Then* π *is a prime element in* $\mathbb{Z}[i]$.

Proof. It suffices to check that π is an irreducible element by Theorems 3.1.11
and 3.5.7 and Proposition 3.5.3. Assume that $\pi = ab$. Then $N(\pi) = N(a)N(b)$.
This means that $N(a) = p$ or $N(b) = p$. If for example $N(a) = p$ then $N(b) = 1$ and b is a unit (see Example 3.1.5). So π is irreducible.   □

We have indicated in the proof of Theorem 3.1.11 that $\mathbb{Z}[i]$ is a Euclidean
domain. Let us give some more details. Given $x \in \mathbb{Z}[i]$ and $d \in \mathbb{Z}[i] \setminus \{0\}$, we
can form $x/d \in \mathbb{Q}[i]$.

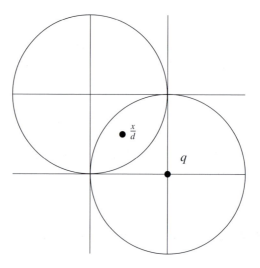

The above picture shows that we may find $q = q_1 + iq_2 \in \mathbb{Z}[i]$ such that

$$\left|\frac{x}{d} - q\right|^2 < 1.$$

Multiplying both sides by $N(d)$ and using $N(ab) = N(a)N(b)$ we get $N(x - qd) < N(d)$, showing that $\mathbb{Z}[i]$ is a Euclidean domain and hence a principal ideal domain and a unique factorization domain. Let us dig a little deeper into the prime elements in $\mathbb{Z}[i]$. We wish to prove that prime numbers congruent to 1 modulo 4 fail to be prime elements in $\mathbb{Z}[i]$. This agrees with the examples $5 = (2 + i)(2 - i)$ and $13 = (3 + 2i)(3 - 2i)$. First, a classical result that deserves to be singled out:

Lemma 3.5.12 (Lagrange) *Let p be a prime number. If $p \equiv 1 \pmod 4$ then the congruence*

$$x^2 \equiv -1 \pmod p$$

can be solved by $x = (2n)!$ where $p = 4n + 1$.

Proof. This is a consequence of Wilson's theorem, which says that $(4n)! \equiv -1 \pmod p$ (see Exercise 1.29(ii)). Write $(4n)!$ as

$$4n(4n - 1) \cdots (4n - 2n + 1) \, 2n(2n - 1) \cdots 2 \cdot 1.$$

Since $4n \equiv -1 \pmod p$, $4n - 1 \equiv -2 \pmod p$, ..., $4n - 2n + 1 \equiv -2n \pmod p$ it follows that $-1 \equiv (4n)! \equiv ((2n)!)^2$. Thus $x = (2n)!$ solves the congruence. $\square$

Remark 3.5.13 There is another proof of Lemma 3.5.12, which in a way is simpler. It also suggests an effective algorithm for computing a solution to the congruence $x^2 \equiv -1 \pmod p$. Suppose that a is a quadratic non-residue modulo p (see Section 1.11). Then we know by Theorem 1.11.4 that

$$a^{(p-1)/2} \equiv -1 \pmod p.$$

So when $p \equiv 1 \pmod 4$, $x = a^{(p-1)/4}$ (or its remainder $[a^{(p-1)/4}]_p$, which can be computed effectively using repeated squaring) is a solution to $x^2 \equiv -1 \pmod p$.

Corollary 3.5.14 *A prime number $p \equiv 1 \pmod 4$ is not a prime element in $\mathbb{Z}[i]$.*

Proof. By Lemma 3.5.12 we can find an integer x such that $x^2 \equiv -1$ (mod p). Then $p \mid x^2 + 1 = (x + i)(x - i)$. But $p \nmid x + i$ and $p \nmid x - i$, since $x/p + (1/p)i \notin \mathbb{Z}[i]$ and $x/p - (1/p)i \notin \mathbb{Z}[i]$. This shows that p is not a prime element in $\mathbb{Z}[i]$.   □

Let us move on to prove Fermat's famous two-square theorem.

Theorem 3.5.15 (Fermat) *A prime number $p \equiv 1$ (mod 4) is a sum of two uniquely determined squares.*

Proof. Assume that $p = a^2 + b^2$ for some integers $a, b \in \mathbb{Z}$. Then $x = a + bi$ is an element of $\mathbb{Z}[i]$ with $N(x) = p$. So x is a prime element by Proposition 3.5.11. If $p = c^2 + d^2$ for some other integers $c, d \in \mathbb{Z}$ then $p = (c + id)(c - id) = (a + bi)(a - bi)$ gives two irreducible factorizations of p. Now the uniqueness of the squares can be deduced from the fact that p has a unique irreducible factorization and $\mathbb{Z}[i]^* = \{1, -1, i, -i\}$. For example, if $c + di \mid a + bi$ then $a + bi = u(c + di)$, where $u \in \mathbb{Z}[i]^*$. The four units correspond to the following cases: $c = a, d = b; c = -a, d = -b; c = b, d = -a;$ $c = -b, d = -a$. These cases show that the squares are unique.

For the existence we need to prove that a prime number $p \equiv 1$ (mod 4) is a sum of two squares. We know by Corollary 3.5.14 that p is not a prime element in $\mathbb{Z}[i]$. Let $\pi = a + bi \in \mathbb{Z}[i]$ be a prime element such that $p = \pi x$, where $x \in \mathbb{Z}[i]$. Then x is not a unit (if x were a unit then p would be a prime element) and so $N(x) > 1$. This means that $N(\pi) = p$, since $p^2 = N(p) = N(\pi)N(x)$. But then $N(\pi) = \pi \bar{\pi} = a^2 + b^2 = p$ and we have expressed p as a sum of two squares.   □

3.5.6 The Euclidean algorithm strikes again

We have proved that every prime number p congruent to 1 modulo 4 is a sum of two squares. So far, trial and error has enabled us to guess identities like $5 = 1^2 + 2^2$, $13 = 3^2 + 2^2$, ... There is a beautiful algorithm (due to Cornacchia, based on a continued-fractions algorithm due to Serret and Hermite) for finding the two squares that sum up to p.

We will describe the algorithm but leave out the proof that it works (see [25] for the proof or do Exercise 3.40 (HOF) on your own – the title of this subsection is the title of [25]). The key point is to compute $x \in \mathbb{N}$ such that $x^2 \equiv -1$ (mod p). This can be done effectively using Remark 3.5.13. Pick a number $a = 1, 2, \ldots, p - 1$ at random. Since the numbers of quadratic residues and quadratic non-residues modulo p are the same, the probability that a is a quadratic non-residue is $1/2$. If this is so then $a^{(p-1)/2} \equiv -1$ and Remark 3.5.13 gives the solution. If not, try another random a. The probability of not having encountered a quadratic non-residue after n trials is $(1/2)^n$.

Suppose that x is a solution to the congruence $x^2 \equiv -1 \pmod{p}$. We may assume that $0 < x < p/2$ (why?). Then use the Euclidean algorithm on p and x. The first two remainders $a, b < \sqrt{p}$ satisfy $p = a^2 + b^2$. Here are some examples.

Example 3.5.16 Let $p = 41$. Then $x = 9$ satisfies $x^2 \equiv -1 \pmod{p}$. Let us apply the Euclidean algorithm to 41 and 9 (see Example 1.5.5).

i	-1	0	1	2	3	4
r_i	41	9	5	4	1	0
q_i			4	1	1	4
a_i	1	0	1	-1	2	-9
b_i	0	1	-4	5	-9	41

The first two remainders $< \sqrt{41}$ are 5 and 4, and $41 = 5^2 + 4^2$.

Example 3.5.17 Let $p = 113$. Then $x = 15$ satisfies $x^2 \equiv -1 \pmod{p}$. Let us apply the Euclidean algorithm to 113 and 15:

i	-1	0	1	2	3	4
r_i	113	15	8	7	1	0
q_i			7	1	1	7
a_i	1	0	1	-1	2	-15
b_i	0	1	-7	8	-15	113

The first two remainders $< \sqrt{113}$ are 8 and 7, and $113 = 8^2 + 7^2$.

There are many patterns in the above examples. If you look at the row with remainders then it appears backwards, up to a sign in the bottom row. Notice also that the algorithm seems to stop after an even number of steps n and that $p \mid r_j^2 + r_{n-j-2}^2$ for $j = -2, -1, 0, \ldots, n$. These facts and the complete proof that the algorithm works can be found in [25].

3.5.7 Prime numbers congruent to 1 modulo 4

Euclid proved that there are infinitely many prime numbers. His proof can be extended to the stronger statement that there are infinitely many prime numbers congruent to 1 modulo 4. Here we show how the Gaussian integers help us in proving this statement.

Lemma 3.5.18   *A prime number $p \equiv 3 \pmod 4$ is a prime element in $\mathbb{Z}[i]$.*

Proof.   Let $\pi = c + id \in \mathbb{Z}[i]$ be a prime element dividing p. Write this as $p = \pi x$ for $x \in \mathbb{Z}[i]$. Then $N(\pi)N(x) = N(p) = p^2$. Thus $N(\pi) = p$ or $N(\pi) = p^2$. If $N(\pi) = \pi \bar{\pi} = c^2 + d^2 = p$ then p is the sum of two squares, but the sum of two squares is not congruent to 3 modulo 4 (see Exercise 3.32). So we must have $N(\pi) = p^2$. Therefore $N(x) = 1$, x is a unit and p is a prime element, since it is a unit times a prime element. $\square$

Corollary 3.5.19   *If p is an odd prime number dividing $x^2 + 1$ for some $x \in \mathbb{Z}$ then $p \equiv 1 \pmod 4$.*

Proof.   Let p be a prime number dividing $x^2 + 1$. If $p \equiv 3 \pmod 4$ then p is a prime element in $\mathbb{Z}[i]$. Thus $p \mid (x^2 + 1) = (x + i)(x - i)$, but p does not divide either $x + i$ or $x - i$. So we must have $p \equiv 1 \pmod 4$. $\square$

Theorem 3.5.20   *There are infinitely many primes congruent to 1 modulo 4.*

Proof.   Suppose there are only finitely many prime numbers $q_1, \ldots, q_s$ congruent to 1 modulo 4. Then form the number

$$N = (q_1 q_2 \cdots q_s)^2 + 1.$$

By Corollary 3.5.19, N is divisible by a prime $p \equiv 1 \pmod 4$. But $p \notin \{q_1, \ldots, q_s\}$, since $q_i \nmid N$ for $i = 1, \ldots, s$. $\square$

Remark 3.5.21   A celebrated result of Dirichlet (1805–59) states that an arithmetic progression

$$b, \quad b + a, \quad b + 2a, \quad b + 3a, \quad \ldots$$

contains infinitely many primes if a and b are relatively prime. It is one of the truly deep theorems of number theory (we have just seen that the case $b = 1$ and $a = 4$ is not particularly easy). We will prove that there are infinitely many primes $\equiv 1 \pmod n$ for every $n \geq 2$ after having introduced cyclotomic polynomials in Chapter 4.

3.5.8 Fermat's last theorem

Suppose we wish to prove Fermat's last theorem (FLT) for $n = 3$: $x^3 + y^3 = z^3$ has no solutions, assuming that x, y and z are non-zero natural numbers. A very fruitful idea is to view the identity $x^3 + y^3 = z^3$ in a ring containing not only the integers but also complex numbers! In fact putting $\omega = e^{2\pi i/3} = -\frac{1}{2} + \frac{\sqrt{3}}{2}i$, we have $\omega^2 + \omega + 1 = 0$ and

$$x^3 + y^3 = (x + y)(x + \omega y)(x + \omega^2 y). \tag{3.3}$$

This factorization does not make sense in the ring $\mathbb{Z}$, but in the enlarged ring $\mathbb{Z}[\omega] = \{x + y\omega \mid x, y \in \mathbb{Z}\}$.

One can prove that $\mathbb{Z}[\omega]$ is a unique factorization domain. A further analysis [12] of the identity (3.3) in the ring $\mathbb{Z}[\omega]$ (using the prime element $1 - \omega$) proves FLT for $n = 3$. For any odd prime number p we have the factorization

$$x^p + y^p = (x + y)(x + \omega y)\cdots(x + \omega^{p-1}y)$$

in the ring $\mathbb{Z}[\omega] = \{a_0 + a_1\omega + \cdots + a_{p-2}\omega^{p-2} \mid a_0, \ldots, a_{p-2} \in \mathbb{Z}\}$, where $\omega = e^{2\pi i/p}$. In 1847 Lamé (1795–1870) announced to the French academy that he had proved FLT. His "proof" was based on the (wrong) assumption that $\mathbb{Z}[\omega]$ is a unique factorization domain for all primes p. A letter from Kummer (1810–93) pointed out the mistake and introduced "ideal complex numbers" to restore unique factorization. Kummer proved the remarkable theorem that FLT holds for an odd prime number p if p does not divide the numerator of any of the Bernoulli numbers $B_2, B_4, \ldots, B_{p-3}$ (such a prime number is called regular). The Bernoulli numbers are given by the coefficients B_n in the power series expansion

$$\frac{x}{e^x - 1} = \sum_{n=0}^{\infty} B_n \frac{x^n}{n!}$$

and can be computed using a variant of Newton's method ([5], Section 4.4). The first few Bernoulli numbers are

$$B_0 = 1, \quad B_1 = -\tfrac{1}{2}, \quad B_2 = \tfrac{1}{6}, \quad B_3 = 0, \quad B_4 = -\tfrac{1}{30}, \quad B_5 = 0, \quad B_6 = \tfrac{1}{42}.$$

Using Kummer's result we see that FLT holds for $p = 3, 5, 7$. The thirty-second Bernoulli number is

$$-\frac{7709321041217}{510}.$$

Since $7709321041217 = 37 \cdot 683 \cdot 305065927$, this shows that 37 is an irregular prime number (in fact the first). The Danish mathematician J. L. W. Jensen (1859–1925) [15] showed in 1915 that there are infinitely many irregular prime numbers. Analyzing irregular prime numbers, FLT was proved by S. Wagstaff for all n up to 125000 in 1978.

Kummer's insights led to an immense amount of important mathematics. FLT was finally proved in the early morning (EDST) of September 19, 1994 by the British mathematician Andrew Wiles of Princeton University. Wiles' proof [26] utilizes the most advanced techniques of modern mathematics and builds heavily on results obtained in the late twentieth century.

3.6 Exercises

1. Show that a zero divisor cannot be a unit.
2. We may view the complex numbers $\mathbb{C}$ as the real plane $\mathbb{R}^2$ with basis 1 and i. This means that the real plane as an abelian group can be equipped with a multiplication making it into a field. Can we extend this multiplication to obtain a ring multiplication on $\mathbb{R}^3$? View $\mathbb{R}^3$ as $a + bi + cj$, where $a, b, c \in \mathbb{R}$. Suppose that we have a multiplication on $\mathbb{R}^3$, making it into a ring, such that $ii = i^2 = -1$. Then $ij = x + yi + zj$ for $x, y, z \in \mathbb{R}$. Multiply both sides of this equation by i to show that such a multiplication cannot exist (however, if you add one more dimension then you can obtain a multiplication, as we saw in Example 3.1.2).
3. Let R be a ring. Prove that $0 \cdot x = 0$ and $-x = (-1) \cdot x$ for every $x \in R$.
4. Prove that an ideal I in a ring R is the whole ring if and only if $1 \in I$.
5. Let R be a ring and $r_1, \dots, r_n \in R$. Prove that the subset
   $\langle r_1, \dots, r_n \rangle = \{\lambda_1 r_1 + \cdots + \lambda_n r_n \mid \lambda_1, \dots, \lambda_n \in R\}$ is an ideal in R.
6. Let I be an ideal in a ring R. Prove that $\langle r_1, \dots, r_n \rangle \subseteq I$ if $r_1, \dots, r_n \in I$.
7. Let M be a subset of a ring R. Prove that $\langle f \mid f \in M \rangle$ (see Remark 3.1.6) is an ideal.
8. Let I and J be ideals in the ring R.
   (i) Prove that

   $$I \cap J$$

   is an ideal in R.
   (ii) Prove that

   $$I + J = \{a + b \mid a \in I, \ b \in J\}$$

   is an ideal in R.

(iii) Prove that

$$IJ = \left\{ \sum_{i=1}^{n} a_i b_i \mid n \geq 1,\ a_i \in I,\ b_i \in J \right\}$$

is an ideal in R.

(iv) Prove that $IJ \subseteq I \cap J$. Give an example where $IJ \subsetneq I \cap J$.

(v) Is $\{ab \mid a \in I,\ b \in J\}$ an ideal in R?

9. Let

$$I_1 \subseteq I_2 \subseteq I_3 \subseteq \cdots$$

be an increasing sequence of ideals in a ring R. Prove that the union of the ideals is an ideal. Give an example of two ideals I and J such that $I \cup J$ is not an ideal.

10. Let R be a ring with the property that every ideal $I \subseteq R$ is finitely generated i.e. there are finitely many elements $r_1, \ldots, r_n \in R$ such that $I = \langle r_1, \ldots, r_n \rangle$ (such a ring is called *noetherian*).

(i) Prove that an increasing sequence (chain) of ideals

$$I_1 \subseteq I_2 \subseteq I_3 \subseteq \cdots$$

must stabilize, i.e. there is a natural number N such that
$I_N = I_{N+1} = \ldots$

(ii) Is an ideal $J \neq R$ in a noetherian ring contained in a maximal ideal?

11. Prove that the kernel $\operatorname{Ker} f = \{r \in R \mid f(r) = 0\} \subseteq R$ of a ring homomorphism $f : R \to S$ is an ideal of R and that the image $f(R)$ is a subring of S.

12. (i) Find integers $\lambda, \mu \in \mathbb{Z}$ such that

$$49\lambda + 13\mu = 1,$$

and show using this that the coset $[13]$ is a unit in $\mathbb{Z}/49\mathbb{Z}$.

(ii) In the following R will denote $\mathbb{Z}/p^l\mathbb{Z}$, where p is a prime and $l > 0$ a natural number. Show that R is not a domain if $l > 1$.

(iii) Show that the number of non-units in R is p^{l-1}.

(iv) Suppose that $r^2 = r$ where $r \in R$. Show that $r = [0]$ or $r = [1]$.

13. Show that the group of units $\mathbb{Z}[i]^*$ in $\mathbb{Z}[i]$ is isomorphic to $\mathbb{Z}/4\mathbb{Z}$.

14. Let $\omega = e^{2\pi i / p}$, where $p \in \mathbb{N}$ and $p > 1$. Prove that

$$\mathbb{Z}[\omega] = \{a_0 + a_1\omega + \cdots + a_{p-2}\omega^{p-2} \mid a_0, \ldots, a_{p-2} \in \mathbb{Z}\}$$

is a subring of $\mathbb{C}$.

15. (i) Show that $\mathbb{Z}[\sqrt{2}] = \{a + b\sqrt{2} \mid a, b \in \mathbb{Z}\}$ is a subring of $\mathbb{R}$.

   (ii) Show that $\mathbb{Z}[\sqrt{2}]^*$ is infinite (hint: consider powers of $1 + \sqrt{2}$).

16. Let R denote the ring $\mathbb{Z}[i]/\langle 1 + 3i \rangle$.

   (i) Show that $i - 3 \in \langle 1 + 3i \rangle$ and that $[i] = [3]$ in R. Use this to prove that $[10] = [0]$ in R and that $[a + bi] = [a + 3b]$, where $a, b \in \mathbb{Z}$.

   (ii) Show that the unique ring homomorphism

$$\varphi : \mathbb{Z} \to R$$

   is surjective.

   (iii) Show that $1 + 3i$ is not a unit and that $1 + 3i$ does not divide 2 and 5 in $\mathbb{Z}[i]$. Conclude that $\operatorname{Ker} \varphi = 10\mathbb{Z}$.

   (iv) Show that $R \cong \mathbb{Z}/10\mathbb{Z}$.

17. Let R be a commutative ring and let I, J, where $I \subseteq J$, be ideals in R.

   (i) Show that $\varphi : R/I \to R/J$ given by $\varphi(x + I) = x + J$ is a well defined, surjective ring homomorphism.

   (ii) Let $R = \mathbb{Z}[i]$. Consider $n \in \mathbb{Z} \setminus \{0\}$ and the ideal $I = Rn$ in R. Show that $a + bi \in I$ if and only if $n|a$ and $n|b$. Show that R/I is a finite ring.

   (iii) Use the notation from (ii). Let $J \neq 0$ be an ideal in R. Show that $J \cap \mathbb{Z}$ is an ideal in $\mathbb{Z}$ and that $J \cap \mathbb{Z} \neq 0$.

   (iv) Use the notation from (ii) and (iii). Show that R/J is a finite ring.

18. Prove that a ring having characteristic zero contains a subring isomorphic to $\mathbb{Z}$.

19. Let $\varphi : R \to S$ be a ring homomorphism. Show that $J = \operatorname{Ker}(\varphi)$ is a prime ideal if S is a domain. Show that J is a maximal ideal if S is a field and φ is surjective.

20. Let I be an ideal in the ring R and let $\pi : R \to R/I$ denote the canonical ring homomorphism.

   (i) Let $J \subseteq R/I$ be an ideal. Prove that $\pi^{-1}(J)$ is an ideal containing I.

   (ii) Let $I' \supseteq I$ be an ideal containing I. Prove that $\pi(I')$ is an ideal in R/I.

   (iii) Prove that π and π^{-1} give a one to one correspondence, preserving $\subseteq$, between ideals in R containing I and ideals in R/I. Use this to prove that R/I is a field if and only if I is a maximal ideal.

   (iv) List the (finitely many) ideals in $\mathbb{Z}/24\mathbb{Z}$.

21. Let R be a non-zero commutative ring. Prove that R/P is a domain if P is a prime ideal.

22. Let I and J be ideals and P a prime ideal of R. Prove that if $IJ \subseteq P$ then $I \subseteq P$ or $J \subseteq P$.

23. Prove that a finite domain F is a field (hint: consider $x \in F \setminus \{0\}$ along with $x^2, x^3, \ldots$).
24. What is the fraction field of a field?
25. Prove that every ideal in the quotient ring R/I of a principal ideal domain R is principal. Give an example of a ring which is not a domain but for which every ideal is a principal ideal.
26. What are the units in $\mathbb{Z}/8\mathbb{Z}$? Give an example of a ring R with an element $x \neq 0, 1$ such that $x^2 = x$. Is R a domain? Suppose that every $x \in R$ satisfies $x^2 = x$. Show that char $R = 2$.
27. Let $f(z) = \bar{z}$ denote the conjugation map for a complex number $z \in \mathbb{C}$. Prove that f is a ring homomorphism $\mathbb{Z}[i] \to \mathbb{Z}[i]$ and that $f(\pi)$ is a prime element if $\pi \in \mathbb{Z}[i]$ is a prime element.
28. Is the remainder r in the definition of a Euclidean domain unique?
29. Compute a greatest common divisor d of $a = 4 + 5i$ and $b = 7 + 8i$ in $\mathbb{Z}[i]$ along with $\lambda, \mu \in \mathbb{Z}[i]$ such that $\lambda a + \mu b = d$.
30. Let $\mathbb{Z}[\omega] = \{x + \omega y \mid x, y \in \mathbb{Z}\}$, where $\omega^2 + \omega + 1 = 0$. Let $z = x + \omega y \in \mathbb{Z}[\omega]$ and let $\bar{z}$ denote the complex conjugate of z.
    (i) Prove that $N(z) = z\bar{z} = x^2 - xy + y^2$ and that $N(z_1 z_2) = N(z_1)N(z_2)$. Show that $z \in \mathbb{Z}[\omega]$ is a unit if and only if $N(z) = 1$.
    (ii) Prove that $z \in \mathbb{Z}[\omega]$ is irreducible if $N(z)$ is a prime number.
    (iii) Prove that $\mathbb{Z}[\omega]$ is a Euclidean domain.
    (iv) Prove that $1 - \omega$ is a prime element in $\mathbb{Z}[\omega]$.
31. Is $\mathbb{Z}[\sqrt{-3}] = \{x + y\sqrt{-3} \mid x, y \in \mathbb{Z}\}$ a Euclidean ring?
32. Prove that the square of a number is either $\equiv 0 \pmod 4$ or $\equiv 1 \pmod 4$.
33. Let π denote a prime element in $\mathbb{Z}[i]$ such that $\pi \notin \mathbb{Z}, i\mathbb{Z}$. Prove that $N(\pi) = 2$ or $N(\pi) = p$, where p is a prime number $\equiv 1 \pmod 4$. Give a complete classification of the prime elements in $\mathbb{Z}[i]$ using the prime numbers in $\mathbb{Z}$.
34. Prove that there are infinitely many prime numbers $\equiv 3 \pmod 4$ by imitating the proof of Theorem 1.8.2 with $N = 4p_1 \cdots p_n - 1$.
35. How do you write 221 as a sum of two squares using that $17 = 1^2 + 4^2$ and $13 = 2^2 + 3^2$?
36. Show that 51 is not a sum of two squares.
37. Write 137 as a sum of two squares using the algorithm outlined in subsection 3.5.6.
38. How do the points shown in the following diagram relate to the Gaussian integers?

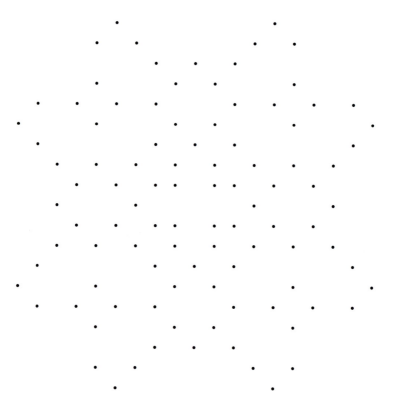

39. Let p be a prime number. Define

$$\mathbb{Z}_{(p)} = \left\{ \frac{a}{s} \in \mathbb{Q} \mid p \nmid s \right\} \subseteq \mathbb{Q}.$$

  (i) Prove that $\mathbb{Z}$ is a subring of $\mathbb{Z}_{(p)}$ and that $\mathbb{Z}_{(p)}$ is a subring of $\mathbb{Q}$. Show that the field of fractions of $\mathbb{Z}_{(p)}$ is isomorphic to $\mathbb{Q}$.

  (ii) Find the units $\mathbb{Z}_{(p)}^*$.

  (iii) Show that every non-zero element $x \in \mathbb{Z}_{(p)}$ can be written uniquely as up^n, where u is a unit and $n \geq 0$.

  (iv) Let I be a non-zero ideal of $\mathbb{Z}_{(p)}$. Show that $I = \langle p^n \rangle$ for some $n \geq 0$.

  (v) Show that $\mathbb{Z}_{(p)}$ contains only one maximal ideal.

40. **(HOF)** Prove that the algorithm given in subsection 3.5.6 works without consulting [25].

41. **(HOF)** Let $R = \mathbb{Z}[\xi] = \{x + y\xi \mid x, y \in \mathbb{Z}\} \subseteq \mathbb{C}$, where $\xi = (1 + \sqrt{-19})/2$. Prove that R is a principal ideal domain that is not a Euclidean domain.

4 Polynomials

The set of functions $f : \mathbb{R} \to \mathbb{R}$ is a ring in a very straightforward manner: the sum of two functions f and g is $(f + g)(x) = f(x) + g(x)$ and the product $(fg)(x) = f(x)g(x)$. The subset $\{a_0 + a_1 x + \cdots + a_n x^n \mid a_i \in \mathbb{R}\}$ of polynomials is a subring of this ring. It is easy to show that the above addition and multiplication lead to the addition

$$(a_0 + a_1 x + a_2 x^2 + \cdots) + (b_0 + b_1 x + b_2 x^2 + \cdots)$$
$$= (a_0 + b_0) + (a_1 + b_1)x + (a_2 + b_2)x^2 + \cdots$$

and the multiplication

$$(a_0 + a_1 x + a_2 x^2 + \cdots)(b_0 + b_1 x + b_2 x^2 + \cdots)$$
$$= (a_0 b_0) + (a_1 b_0 + a_0 b_1)x + (a_2 b_0 + a_1 b_1 + a_0 b_2)x^2 + \cdots$$

of polynomials. The marvelous thing is that this addition and multiplication is algebraic in nature and makes sense even if we replace the coefficients with elements in an arbitrary (commutative) ring.

In many ways polynomials form the heart of algebra. In this chapter we begin by introducing polynomials formally. Straight after the formal introduction we will give a surprising application of the addition and multiplication of polynomials. We will show how one can easily compute the remainder of a binomial coefficient divided by a prime number p by computing with polynomials with coefficients in $\mathbb{F}_p$.

The division algorithm for polynomials is crucial. We give it here in a slightly modified form (see Proposition 4.2.4) to make clear why the general division algorithm in several variables (see Proposition 5.3.1) is really a natural extension. After the important Theorem 4.3.5 we move on to the classical subjects of cyclotomic polynomials and finite fields. You will miss a golden opportunity

if you do not immerse yourself in these topics. It is your ticket to a real understanding of this chapter.

As promised in Section 1.11 we give a proof of Gauss' famous theorem
on quadratic reciprocity. The proof uses Freshman's Dream (Theorem 3.3.9)
and computations with Gauss sums in a suitable quotient ring of a polynomial
ring. You will have the necessary background to learn a proof of a really deep
theorem in number theory just by knowing basic properties of polynomials.

Freshman's Dream is also a key player in the odd fact that there are
fast algorithms for factoring polynomials with coefficients in $\mathbb{F}_p$ into irreducible polynomials. We go through this by describing the basic steps of
Berlekamp's algorithm. Notice the stark contrast with the integers $\mathbb{Z}$, where
no one (so far) has come up with a fast algorithm for factoring. Most of the
mathematics in this chapter can be traced back to the seminal work [11] of
Gauss.

4.1 Polynomial rings

We will introduce the polynomial ring formally. It will be important for us to
view polynomials as purely algebraic objects and not as a subring of a ring of
functions (see Exercise 4.1).

Let R be a ring (commutative as usual) and $R[\mathbb{N}]$ the set of functions
$f : \mathbb{N} \to R$ such that $f(n) = 0$ for $n \gg 0$ (here one should think of the
polynomial $f(0) + f(1)X + f(2)X^2 + \cdots$). Given $f, g \in R[\mathbb{N}]$ we define
their sum as $(f + g)(n) = f(n) + g(n)$. Inspired by the way "real-world
polynomials" multiply, we define

$$(fg)(n) = \sum_{i+j=n} f(i)g(j),$$

where $i, j \in \mathbb{N}$. For example $(fg)(3) = f(3)g(0) + f(2)g(1) + f(1)g(2) +
f(3)g(0)$. We let $X^i \in R[\mathbb{N}]$ denote the function

$$X^i(n) = \begin{cases} 1 & \text{if } n = i, \\ 0 & \text{if } n \neq i. \end{cases}$$

Notice that $X^i X^j = X^{i+j}$, where $i, j \in \mathbb{N}$. We view an element $a \in R$ as the
function

$$a(n) = \begin{cases} a & \text{if } n = 0, \\ 0 & \text{if } n > 0 \end{cases}$$

in $R[\mathbb{N}]$. So, an element $f \in R[\mathbb{N}]$ can be written as

$$f = a_0 + a_1 X + \cdots + a_n X^n,$$

where $a_i = f(i)$ and $f(i) = 0$ if $i > n$. Notice that $1 = X^0$ is (the) neutral element for the multiplication. The neutral element for $+$ is $0 \in R$. Clearly $fg = gf$ and $f(g + h) = fg + fh$ for $f, g, h \in R[\mathbb{N}]$. With these tools it becomes easy (see Exercise 4.2) to verify the associative rule for the multiplication, i.e. $f(gh) = (fg)h$ for every $f, g, h \in R[\mathbb{N}]$, since we may assume that $h = cX^m$, where $c \in R$ and $m \in \mathbb{N}$.

Definition 4.1.1   We define the polynomial ring $R[X]$ in one variable over the ring R as $R[\mathbb{N}]$. Here X denotes the function $X^1 \in R[\mathbb{N}]$. A *term* is a polynomial of the form aX^m, where $a \in R \setminus \{0\}$. A polynomial $f \in R[X]$ can be written

$$a_0 + a_1 X + a_2 X^2 + \cdots + a_n X^n,$$

where $a_0, \ldots, a_n \in R$ are called the *coefficients* of f. If $a_n \neq 0$ we put $\deg(f) = n$ and call a_n the *leading coefficient* of f. In this case $\deg(f)$ is called the *degree* of f and $a_{\deg(f)}X^{\deg(f)}$ its *leading term*. A non-zero polynomial is called *monic* if its leading coefficient is 1.

Remark 4.1.2   The degree of a polynomial is a function $\deg : R[X] \setminus \{0\} \to \mathbb{N}$. It is an extremely useful invariant of a polynomial. The degree of the zero polynomial is not defined.

Now you have seen the formal definition of $R[X]$. When computing with polynomials it pays to treat them as the usual polynomial expressions that we know.

Remark 4.1.3   Two polynomials $f = a_m X^m + \cdots + a_1 X + a_0$ and $g = b_n X^n + \cdots + b_1 X + b_0$ in $R[X]$ are the same if and only if $a_0 = b_0$, $a_1 = b_1, \ldots$. This is clear when we view the polynomials as functions $\mathbb{N} \to R$. Two functions $\mathbb{N} \to R$ are the same if and only if they assume the same value for every $n \in \mathbb{N}$.

We have proved that $R[X]$ really is a ring when R is a ring. This means that all the concepts from Chapter 3 apply. For example, it makes sense to ask whether $R[X]$ is a domain, a Euclidean domain or a unique factorization domain. It also makes sense to ask whether a polynomial in $R[X]$ is a unit or a zero divisor.

4.1.1 Binomial coefficients modulo a prime number

Let us pause for a while after having introduced polynomials formally. We will give an example showing that computations with polynomials can be helpful in reasoning about numbers and congruences. We wish to prove that

$$7 \mid \binom{55}{22}.$$

Judging from the size of the binomial coefficient on the right, this may not be an easy task, unless of course we dig a bit deeper into the structure of the polynomial ring. If n is a natural number > 1 then every $x \in \mathbb{N}$ has a unique n-adic expansion (see Exercise 1.5)

$$x = a_0 + a_1 n + a_2 n^2 + \cdots + a_r n^r,$$

where $r \in \mathbb{N}$, $a_i \in \mathbb{N}$ and $0 \le a_i < n$ for $i = 0, \dots, r$. Recall Freshman's Dream (Theorem 3.3.9) from Chapter 3: if R is a commutative ring of prime characteristic p then

$$(a + b)^{p^r} = a^{p^r} + b^{p^r}$$

for $a, b \in R$ and $r \in \mathbb{N}$. This shows that if m is a natural number with the p-adic expansion

$$m = a_0 + a_1 p + \cdots + a_r p^r$$

then

$$(1 + X)^m = (1 + X)^{a_0}(1 + X^p)^{a_1} \cdots (1 + X^{p^r})^{a_r}$$

in the polynomial ring $\mathbb{F}_p[X]$ (which is a commutative ring of characteristic p). Now let

$$n = b_0 + b_1 p + \cdots + b_s p^s$$

be another natural number and its p-adic expansion. Compare the coefficients of the left hand side of the previous equation,

$$(1 + X)^m = \sum_{n=0}^{m} \binom{m}{n} X^n,$$

with the coefficients of its right hand side,

$$(1 + X)^{a_0}(1 + X^p)^{a_1} \cdots (1 + X^{p^r})^{a_r}$$

$$= \left(\sum_{b_0=0}^{a_0} \binom{a_0}{b_0} X^{b_0} \right) \left(\sum_{b_1=0}^{a_1} \binom{a_1}{b_1} X^{pb_1} \right) \cdots \left(\sum_{b_r=0}^{a_r} \binom{a_r}{b_r} X^{p^r b_r} \right).$$

A term in the product above is given uniquely as the product of a term from the first factor, a term from the second factor and so on. This follows from the uniqueness of the p-adic expansion. Two polynomials are the same if and only if their coefficients are the same (Remark 4.1.3). This leads to the surprising identity

$$\binom{m}{n} \equiv \binom{a_0}{b_0}\binom{a_1}{b_1} \cdots \pmod{p},$$

where

$$\binom{r}{s} = \frac{r(r-1)\cdots(r-s+1)}{s(s-1)\cdots 2 \cdot 1}.$$

Thus $p \mid \binom{m}{n}$ if and only if $a_i < b_i$ for some i. Expanding 7-adically 55 and 22 we get

$$55 = 6 + 1 \cdot 7^2,$$
$$22 = 1 + 3 \cdot 7^1.$$

Thus

$$\binom{55}{22} \equiv \binom{6}{1}\binom{0}{3}\binom{1}{0} = 0 \pmod{7},$$

so 7 divides $\binom{55}{22}$.

4.2 Division of polynomials

We move on to describe the important division algorithm for polynomials. First, we give a few properties of the degree function.

Example 4.2.1 If $R = \mathbb{Z}/4\mathbb{Z}$ and $f = g = 2X + 1$ then $fg = 1$, so that $\deg(fg) = 0$ but $\deg(f) = \deg(g) = 1$. Remember that when we write 2 in the ring $\mathbb{Z}/4\mathbb{Z}$ it really means $[2] = 2 + 4\mathbb{Z}$ (see Remark 3.3.4).

The above example shows that the formula $\deg(fg) = \deg(f) + \deg(g)$ for $f, g \in R[X] \setminus \{0\}$ breaks down in general. It can be repaired by imposing some mild restrictions.

Proposition 4.2.2 *Let $f, g \in R[X] \setminus \{0\}$. If the leading coefficient of f or g is not a zero divisor then*

$$\deg(fg) = \deg(f) + \deg(g).$$

Proof. We may write $f = a_m X^m + \cdots$ and $g = b_n X^n + \cdots$, where a_m, b_n are the leading coefficients (thus $m = \deg(f)$ and $n = \deg(g)$). Then

$$fg = a_m b_n X^{m+n} + \cdots .$$

Since one of a_m and b_n is not a zero divisor, we must have $a_m b_n \neq 0$. Therefore $\deg(fg) = m + n = \deg(f) + \deg(g)$.   □

We have seen in Example 4.2.1 that there can be units in $R[X]$ of degree > 0. This is rather pathological. In most cases units have degree zero. A monic polynomial of degree > 0 can never be a unit (why?).

Proposition 4.2.3 *Let R be a domain. Then $R[X]^* = R^*$.*

Proof. Assume that $f \in R[X]^*$. Then there exists $g \in R[X]$ such that $fg = 1$. Thus $\deg(fg) = \deg(f) + \deg(g) = \deg(1) = 0$ by Proposition 4.2.2. This shows that $\deg(f) = \deg(g) = 0$ and $f, g \in R^* \subseteq R$. Thus $R[X]^* \subseteq R^*$. Clearly $R^* \subseteq R[X]^*$.   □

Now we come to the division algorithm in $R[X]$. It can be viewed as an analogue of division with remainder for the integers (Theorem 1.2.1). We are rephrasing it a little so that it generalizes naturally to the division algorithm for polynomials in several variables later. Notice that $:=$ means assigment to a variable (we use $:=$ to distinguish it from $=$, which has a well defined mathematical meaning).

Proposition 4.2.4 *Let d be a non-zero polynomial in $R[X]$. Assume that the leading coefficient of d is not a zero divisor in R. Given $f \in R[X]$, there exist polynomials $q, r \in R[X]$ such that*

$$f = qd + r$$

and either $r = 0$ or none of the terms in r is divisible by the leading term of d.

Proof. Let aX^m denote the leading term of d, where a is not a zero divisor in R. To begin with we have the identity $f = qd + (r + s)$, where $q = 0$, $r = 0$ and $s = f$. If $s = 0$ we are done. If not, let bX^n denote the leading term of s. If aX^m divides bX^n then $n \geq m$, $b = ca$, for a unique $c \in R$, and $bX^n = cX^{n-m}aX^m$. We put

$$q := q + cX^{n-m},$$
$$s := s - cX^{n-m}d.$$

After these assignments we see that the identity $f = qd + (r + s)$ still holds. If aX^m does not divide bX^n we put

$$r := r + bX^n,$$
$$s := s - bX^n.$$

Again after these assignments the identity $f = qd + (r + s)$ holds. After both assignments r will only contain terms not divisible by the leading term of d. Now proceed with the same steps using the new s. If $s = 0$ the procedure will stop. If not we know that the degree of s has strictly decreased since it does so in both steps. After finitely many steps (the degree of f is finite) we will reach the case $s = 0$.   □

If the leading coefficient of d in Proposition 4.2.4 is invertible then there is a more appealing way of formulating the division of polynomials. This is the content of the following corollary.

Corollary 4.2.5 *Let d be a non-zero polynomial in $R[X]$. Assume that the leading coefficient of d is invertible in R. Given $f \in R[X]$, there exist unique polynomials $q, r \in R[X]$ such that*

$$f = qd + r$$

and either $r = 0$ or $\deg(r) < \deg(d)$.

Proof. An invertible element divides every other element in R. Therefore the leading term of d divides a term of degree n if and only if $\deg(d) \leq n$. In this situation Proposition 4.2.4 may be reformulated as $f = qd + r$, where $r = 0$ or $\deg(r) < \deg(d)$ if $r \neq 0$.

Assume that $f = q_1d + r_1 = q_2d + r_2$, where $q_1, r_1, q_2, r_2 \in R[X]$ and r_1, r_2 satisfy the conditions in the corollary. Then $(q_1 - q_2)d = r_2 - r_1$. If $r_2 - r_1 \neq 0$ then $\deg(q_1 - q_2) + \deg(d) = \deg(r_2 - r_1)$. Since $\deg(r_2 - r_1) \leq \max(\deg(r_1), \deg(r_2))$ (see Exercise 4.3), we get $\deg(d) \leq \deg(r_1)$ or $\deg(d) \leq$

deg(r_2). This is a contradiction. It implies that $r_1 = r_2$ and thereby that $q_1 = q_2$, proving the uniqueness of q and r.   □

The division algorithm for polynomials is illustrated in the following example.

Example 4.2.6   If $f = X^4 + X - 1$ and $d = X - 1$ are polynomials in $\mathbb{Z}[X]$, we may write the algorithm in the proof of Proposition 4.2.4 schematically as

$$
\begin{array}{rcll}
X^4 + X - 1 & : & X - 1 & = \quad X^3 + X^2 + X + 2 \\
\underline{X^4 - X^3} & & & \\
X^3 + X - 1 & & & \\
\underline{X^3 - X^2} & & & \\
X^2 + X - 1 & & & \\
\underline{X^2 - X} & & & \\
2X - 1 & & & \\
\underline{2X - 2} & & & \\
1 & & &
\end{array}
$$

This shows that $X^4 + X - 1 = (X^3 + X^2 + X + 2)(X - 1) + 1$.

Definition 4.2.7   The polynomial r in Corollary 4.2.5 is called the *remainder* of f divided by d.

4.3 Roots of polynomials

The map $j : R \to R[X]$ given by

$$j(r) = r + 0X + 0X^2 + \cdots$$

is an injective ring homomorphism. We identify the image $j(R)$ with R and view R as a subring of $R[X]$ in this way.

Proposition 4.3.1   *Let $f = a_n X^n + \cdots + a_1 X + a_0 \in R[X]$ and $\alpha \in R$. The map $\varphi_\alpha : R[X] \to R$ given by*

$$\varphi_\alpha(f) = f(\alpha) = a_n \alpha^n + \cdots + a_1 \alpha + a_0$$

is a ring homomorphism.

Proof. This follows from the rules for adding and multiplying in $R[X]$. □

This leads us to the crucial concept of a root of a polynomial. Let $f \in R[X]$ be a polynomial. The element $\alpha \in R$ is called a *root* of f if $f(\alpha) = \varphi_\alpha(f) = 0$. We let $V(f) = \{\alpha \in R \mid f(\alpha) = 0\}$ denote the set of roots of $f \in R[X]$. The following corollary is a stepping stone toward introducing the concept of the multiplicity of roots in polynomials.

Corollary 4.3.2 *Let $f \in R[X]$. Then $\alpha \in R$ is a root of f if and only if $X - \alpha$ divides f.*

Proof. Assume α is a root of f. By Corollary 4.2.5 we may write

$$f = q(X - \alpha) + r,$$

where r is a constant ($r \in R$). Substituting α for X on both sides (see Proposition 4.3.1) we get $0 = f(\alpha) = r$, which proves $r = 0$. If $X - \alpha$ divides f then $r = 0$ and α is a root of f. □

If a monic polynomial q divides a non-zero polynomial f then $f = qr$ for a unique r (q is not a zero divisor in $R[X]$) and $\deg(f) = \deg(q) + \deg(r)$ by Proposition 4.2.2. Thus $\deg(q) \leq \deg(f)$. The *multiplicity* of α as a root in a non-zero polynomial f is the largest power $n \in \mathbb{N}$ such that

$$(X - \alpha)^n \mid f.$$

The multiplicity of α in f is denoted $v_\alpha(f)$. Notice that $v_\alpha(f) \leq \deg(f)$ and $f = (X - \alpha)^{v_\alpha(f)} h$, where $h(\alpha) \neq 0$. A *multiple root* in f is a root $\alpha \in R$ with $v_\alpha(f) > 1$.

The following example shows that one needs to exercise some caution with regard to roots. There may be too many of them in pathological cases (see also Exercise 4.6).

Example 4.3.3 Let $R = \mathbb{Z}/6\mathbb{Z}$ and $f = X^2 + 3X + 2 \in R[X]$. Then f can have at most six roots (after all there are only six elements in R). Let us tabulate $f(\alpha)$ for $\alpha \in R$:

α	0	1	2	3	4	5
$f(\alpha)$	2	0	0	2	0	0

We see that $V(f) = \{1, 2, 4, 5\}$. In this case f has four roots but the degree of f is 2. It is not true that $f = (X - 1)(X - 2)(X - 4)(X - 5)$.

The usual type of polynomial $f \in \mathbb{R}[X] \setminus \{0\}$ cannot have more than $\deg(f)$ roots. This is wrong in the general case (Example 4.3.3). However, if R is a domain we can get the "right" bound on the number of roots for a non-zero polynomial in $R[X]$. The following simple lemma captures the essence.

Lemma 4.3.4 *Let R be a domain and $f, g \in R[X]$. Then $V(fg) = V(f) \cup V(g)$.*

Proof. The inclusion $V(fg) \supseteq V(f) \cup V(g)$ is true without any assumptions on R. We will prove that $V(fg) \subseteq V(f) \cup V(g)$. If $\alpha \in V(fg)$ then $(fg)(\alpha) = f(\alpha)g(\alpha) = 0$. Since R is a domain we get $f(\alpha) = 0$ or $g(\alpha) = 0$. Thus $\alpha \in V(f)$ or $\alpha \in V(g)$ and $\alpha \in V(f) \cup V(g)$. □

Theorem 4.3.5 *Let R be a domain and $f \in R[X] \setminus \{0\}$. If $V(f) = \{\alpha_1, \ldots, \alpha_r\}$ then*

$$f = q(X - \alpha_1)^{\nu_{\alpha_1}(f)} \cdots (X - \alpha_r)^{\nu_{\alpha_r}(f)},$$

where $q \in R[X]$ and $V(q) = \emptyset$. The number of roots of f, counted with multiplicity, is bounded by the degree of f.

Proof. We prove this using induction on $\deg(f)$. We will show the induction step and leave the cases $\deg(f) = 0$ and $V(f) = \emptyset$ to the reader. If $\alpha \in V(f)$ then $f = (X - \alpha)^{\nu_\alpha(f)}g$, where $\deg(g) < \deg(f)$ and $g(\alpha) \neq 0$. Thus $V(f) = \{\alpha\} \cup V(g)$ by Lemma 4.3.4 and $\alpha \notin V(g)$. By induction

$$g = q(X - \beta_1)^{\nu_{\beta_1}(g)} \cdots (X - \beta_s)^{\nu_{\beta_s}(g)},$$

where $V(g) = \{\beta_1, \ldots, \beta_s\}$ and $V(q) = \emptyset$. This gives the desired formula

$$f = q(X - \alpha)^{\nu_\alpha(f)}(X - \beta_1)^{\nu_{\beta_1}(f)} \cdots (X - \beta_s)^{\nu_{\beta_s}(f)},$$

where $V(f) = \{\alpha\} \cup V(g) = \{\alpha, \beta_1, \ldots, \beta_s\}$ and $V(q) = \emptyset$. Now it follows by Proposition 4.2.2 that

$$\nu_\alpha(f) + \nu_{\beta_1}(f) + \cdots + \nu_{\beta_s}(f) \leq \deg(f),$$

proving that the number of roots of f counted with multiplicity is bounded by the degree of f. □

As a first example of the usefulness of Theorem 4.3.5 we give a (natural) proof of Wilson's theorem (see Exercise 1.29(ii)), which says that $(p - 1)! \equiv -1 \pmod{p}$ if p is a prime number.

Example 4.3.6   Consider the polynomial $X^p - X \in \mathbb{F}_p[X]$. Then

$$V(X^p - X) = \{0, 1, \ldots, p - 1\}$$

by Fermat's little theorem, Corollary 1.9.2. It follows by Theorem 4.3.5 that

$$X^p - X = qX(X - 1)(X - 2) \cdots (X - (p - 1)),$$

where q is a polynomial of degree zero (which has to be 1 by comparing the leading coefficients on both sides). Comparing coefficients of degree one on the left and right hand sides, we get $1 \cdot 2 \cdots (p - 1) = (p - 1)! = -1$ in $\mathbb{F}_p$. This shows that $(p - 1)! \equiv -1 \,(\mathrm{mod}\ p)$.

We now describe a useful algebraic gadget inspired by differentiation in analysis. We cannot employ the usual definition of the derivative from analysis, so we have to be a little more formal.

4.3.1 Differentiation of polynomials

Let R be a ring and $f = a_n X^n + a_{n-1}X^{n-1} + \cdots + a_1 X + a_0 \in R[X]$. Then

$$D(f) = a_n n X^{n-1} + a_{n-1}(n - 1)X^{n-2} + \cdots + a_1$$

is called the *derivative* of f. When a polynomial is viewed formally as a map $f : \mathbb{N} \to R$ (see Section 4.1), this can be rephrased as $D(f)(n - 1) = nf(n)$ for $n \geq 1$. The following lemma shows that the derivative behaves just as in ordinary differentiation.

Lemma 4.3.7   *Let $f, g \in R[X]$ and $\lambda \in R$. Then*

(i) $D(f + g) = D(f) + D(g)$,
(ii) $D(\lambda f) = \lambda D(f)$,
(iii) $D(fg) = f D(g) + D(f)g$.

Proof. We will prove (iii) and leave (i) and (ii) to the reader. Viewing polynomials formally as maps $\mathbb{N} \to R$, (iii) follows from the identity

$$(f D(g) + D(f)g)(n - 1) = \sum_{i+j=n-1} f(i)D(g)(j) + \sum_{i+j=n-1} D(f)(i)g(j)$$

$$= \sum_{i+j=n-1} f(i)(j + 1)g(j + 1)$$

$$+ \sum_{i+j=n-1} (i + 1)f(i + 1)g(j)$$

$$= \sum_{i+j=n} f(i)jg(j) + \sum_{i+j=n} if(i)g(j)$$

$$= n \sum_{i+j=n} f(i)g(j)$$

$$= n(fg)(n) = D(fg)(n-1),$$

where $n \geq 1$. □

The most useful property of the derivative is the Leibniz rule (Lemma 4.3.7(iii)). We will use the derivative to reason about roots of polynomials, as shown in the lemma below.

Lemma 4.3.8 *Suppose that $f, g \in R[X]$.*

(i) *If $f^2 \mid g$ then $f \mid D(g)$.*
(ii) *An element $\alpha \in R$ is a multiple root of f if and only if α is a root of f and $D(f)$.*

Proof. Assume that $g = qf^2$. Then $D(g) = D(q)f^2 + 2qD(f)f = (D(q) f + 2qD(f))f$ by Lemma 4.3.7(iii). This proves (i). If α is a multiple root of f then $(X - \alpha)^2$ divides f. Therefore $X - \alpha$ divides $D(f)$ by (i) and α is a root of $D(f)$. Now assume that α is a root of f and $D(f)$. Then $f = (X - \alpha)^m h$, where $m = v_\alpha(f) \geq 1$ and $h(\alpha) \neq 0$. If $m = 1$ we get $D(f) = h + (X - \alpha)D(h)$. This leads to $D(f)(\alpha) = h(\alpha) \neq 0$, contradicting that α is a root of $D(f)$. Therefore $m \geq 2$ and α is a multiple root of f. This proves (ii). □

Remark 4.3.9 If the polynomial ring $R[X]$ is of prime characteristic $p > 0$ one encounters many non-constant polynomials with zero derivatives. Take $X^p \in \mathbb{F}_p[X]$ as an example. Here

$$D(X^p) = pX^{p-1} = 0.$$

In fact $D(X^n) = 0$ if and only if p divides n when $X^n \in \mathbb{F}_p[X]$. This looks strange but can be very useful.

4.4 Cyclotomic polynomials

A complex number ξ is called an nth root of unity for a positive integer n if $\xi^n = 1$. Writing ξ in polar coordinates as $re^{i\theta} = r(\cos\theta + i\sin\theta)$, it follows that $r = 1$ and $\theta = k2\pi i/n$ for $k = 0, \ldots, n - 1$ if ξ is an nth root of unity. Of course, n may not be the smallest positive integer with the property $\xi^n = 1$

(if $\xi = i$ then, $\xi^8 = 1$ but already $\xi^4 = 1$). A complex number ζ is called a primitive nth root of unity if $\zeta^n = 1$ and

$$\zeta, \zeta^2, \ldots, \zeta^{n-1} \neq 1,$$

where $n \geq 1$. The eighth roots of unity are plotted below as dots on the unit circle in the complex plane. The bigger dots represent the primitive eighth roots of unity.

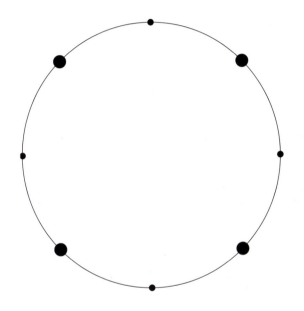

Lemma 4.4.1   *A complex number ζ is a primitive nth root of unity if and only if*

$$\zeta = e^{k2\pi i/n},$$

where $1 \leq k \leq n$ and $\gcd(k, n) = 1$. If ζ is a primitive nth root of unity and $\zeta^m = 1$ then $n \mid m$.

Proof.   The nth roots of unity are $e^{k2\pi i/n}$, where $k = 1, \ldots, n$. Let $\xi = e^{k2\pi i/n}$ be an nth root of unity. If $\xi^m = 1$ then $mk2\pi/n$ is an integer multiple of 2π and therefore $n \mid mk$. Assume that $\gcd(k, n) = 1$. Then $n \mid km$ implies that $n \mid m$ by Corollary 1.5.10. Thus $\xi, \xi^2, \ldots, \xi^{n-1} \neq 1$. Therefore ξ is a primitive nth root of unity. However, if $\gcd(k, n) = g > 1$ then $\xi^{n/g} = 1$ and ξ cannot be a primitive nth root of unity. If ζ is a primitive nth root of unity and $\zeta^m = 1$ then we may write $m = qn + r$, where $0 \leq r < n$. This shows that $\zeta^m = \zeta^r$ and therefore that $r = 0$. □

The set of all nth roots of unity is a subgroup of $\mathbb{C}^*$. This subgroup is isomorphic to the cyclic group $\mathbb{Z}/n\mathbb{Z}$. Using this and Lemma 4.4.1 you get a different angle (see Exercise 4.14) on Proposition 2.7.4 that is more in the spirit of Gauss.

Now we construct a polynomial in $\mathbb{C}[X]$ whose roots are all the primitive nth roots of unity. Although the reason will not yet be clear, this will lead to some amazing algebra later.

Definition 4.4.2 Let $n \in \mathbb{N}$ with $n \geq 1$. The nth cyclotomic polynomial is defined as the polynomial

$$\Phi_n(X) = \prod_{1 \leq k \leq n,\ \gcd(k,n)=1} (X - e^{2\pi ik/n})$$

in $\mathbb{C}[X]$.

Notice that $\deg \Phi_n = \varphi(n)$. The first four cyclotomic polynomials are

$$\Phi_1(X) = X - 1,$$
$$\Phi_2(X) = X + 1,$$
$$\Phi_3(X) = X^2 + X + 1,$$
$$\Phi_4(X) = X^2 + 1.$$

Cyclotomic polynomials are quite complicated. In one version of a manual for the computer algebra system Maple ([21], p. 242, **numtheory[cyclotomic] (n, var)**), it is stated that their coefficients are always ± 1. It appears to be so, when looking at the first 104 cyclotomic polynomials. But

$$\begin{aligned}
\Phi_{105}(X) = {} & 1 + X + X^2 - X^5 - X^6 - 2X^7 - X^8 - X^9 + X^{12} \\
& + X^{13} + X^{14} + X^{15} + X^{16} + X^{17} - X^{20} - X^{22} - X^{24} - X^{26} \\
& - X^{28} + X^{31} + X^{32} + X^{33} + X^{34} + X^{35} + X^{36} - X^{39} - X^{40} \\
& - 2X^{41} - X^{42} - X^{43} + X^{46} + X^{47} + X^{48}
\end{aligned}$$

where the coefficients of X^7 and X^{41} are both -2. I. Schur (1875–1941) proved that the coefficients of Φ_n are unbounded when n goes to infinity. In fact the coefficients of Φ_n have attracted the attention of researchers all through the twentieth century. The coefficients of Φ_n are always $= \pm 1$ if n is a product of two distinct prime numbers (notice that $105 = 3 \cdot 5 \cdot 7$). Cyclotomic polynomials have integer coefficients even though they are defined using roots of unity in the complex plane. This follows from a crucial identity, which turns out to make sense for polynomials over any ring, not just those with complex coefficients.

Proposition 4.4.3 *Let $n \geq 1$. Then*

(i) $X^n - 1 = \prod_{d|n} \Phi_d(X)$;
(ii) *the cyclotomic polynomials have integer coefficients,*

$$\Phi_n(X) \in \mathbb{Z}[X].$$

Proof. The roots of the polynomial on the right hand side of the identity in (i) are the primitive dth roots of unity, where $d \mid n$. They are also roots of the polynomial on the left hand side. However, if $\xi = e^{k2\pi i/n}$, where $1 \leq k \leq n$ is a root of the polynomial on the left hand side, then ξ is a primitive dth root of unity for some $d \leq n$. But $\xi^n = 1$ implies that $d \mid n$ by Lemma 4.4.1, so that ξ is also a root of the polynomial on the right hand side. Thus the polynomials on the left and right hand sides have the same roots. Since they are both monic and neither has multiple roots, they must be identical by Theorem 4.3.5. To prove that $\Phi_n \in \mathbb{Z}[X]$, we use induction. Clearly $\Phi_1 = X - 1 \in \mathbb{Z}[X]$. Let $n > 1$ and $f = \prod_{d<n, \, d|n} \Phi_d$. Then

$$X^n - 1 = \Phi_n f.$$

By induction, f is a monic polynomial in $\mathbb{Z}[X]$. Division of polynomials (Corollary 4.2.4) gives $X^n - 1 = \varphi f + r$, where $r = 0$ or $r \neq 0$, $\deg(r) < \deg(f)$ and $\varphi \in \mathbb{Z}[X]$. The uniqueness of q and r in Corollary 4.2.5 applied inside $\mathbb{C}[X]$ to f and $X^n - 1$ shows that $\Phi_n = \varphi$ and $r = 0$. Thus $\Phi_n = \varphi \in \mathbb{Z}[X]$. □

Now let R be a ring. The unique ring homomorphism $\kappa : \mathbb{Z} \to R$ (see Lemma 3.3.3) gives a ring homomorphism $\kappa' : \mathbb{Z}[X] \to R[X]$ (see Exercise 4.15). In this way we may view the cyclotomic polynomial $\Phi_n \in \mathbb{Z}[X]$ as the polynomial $\kappa'(\Phi_n) \in R[X]$. This leads to the important identity

$$X^n - 1 = \prod_{d|n} \Phi_d(X) \tag{4.1}$$

in $R[X]$ by applying the ring homomorphism κ' to the corresponding identity in $\mathbb{Z}[X]$ (which comes from Proposition 4.4.3).

4.5 Primitive roots

The definition of a primitive root makes sense not only in the complex numbers but also in an arbitrary ring. Notice again that we take a classical idea (from complex numbers) and bring it to use (in fact a great deal of use) in abstract algebra.

Definition 4.5.1   Let R be a ring and n a positive natural number. An element $\alpha \in R$ is called a *primitive nth root of unity* in R if $\alpha^n = 1$ and

$$\alpha, \alpha^2, \ldots, \alpha^{n-1} \neq 1.$$

This leads to the following important lemma.

Lemma 4.5.2   *Let α be an element in a domain R. If $\Phi_n(\alpha) = 0$ and α is not a multiple root of $X^n - 1 \in R[X]$ then α is a primitive nth root of unity in R.*

Proof.   The identity (4.1) in $R[X]$ gives $f\Phi_n = X^n - 1$ for $f \in R[X]$. Therefore $\alpha^n - 1 = f(\alpha)\Phi_n(\alpha) = 0$ and $\alpha^n = 1$. If α is a primitive dth root of unity for $1 \leq d < n$ then $d \mid n$, as in the proof of Lemma 4.4.1. In this case $X^d - 1 = \prod_{e \mid d} \Phi_e(X)$, again by (4.1). Since R is a domain we must have $\Phi_e(\alpha) = 0$ for some $e \mid d$. Therefore α is a root in Φ_n and Φ_e, where $e \mid n$ and $e < n$. This proves by (4.1) that α is a multiple root of $X^n - 1$, contradicting our assumption.   □

Having introduced primitive roots in rings and proved Lemma 4.5.2 we can obtain a simple proof of the following beautiful result due to Gauss.

Theorem 4.5.3 (Gauss)   *Let F be a field and $G \subseteq F^*$ a finite subgroup of the group of the units in F. Then G is cyclic.*

Proof.   Let $N = |G|$ and consider the polynomial

$$X^N - 1 = \prod_{d \mid N} \Phi_d \in F[X].$$

The roots of the polynomial on the left hand side are precisely the elements of G, since $\alpha^N = 1$ for every $\alpha \in G$ by Proposition 2.6.3(ii). There can be no more than N roots by Theorem 4.3.5 and none of these is a multiple root. This shows that Φ_N must have $\deg \Phi_N = \varphi(N)$ roots. These are primitive Nth roots of unity in F by Lemma 4.5.2 and thereby generators of G.   □

Theorem 4.5.3 shows in particular that $\mathbb{F}_p^*$ is a cyclic group, where p is a prime number. An integer a such that $[a]$ generates $\mathbb{F}_p^*$ is called a *primitive root* modulo p. Thus a primitive root a satisfies

$$\mathbb{F}_p^* = \{[1], [a], [a^2], \ldots, [a^{p-2}]\}.$$

If $p = 13$ and $a = 2$ we have

$$\mathbb{F}_{13}^* = \{[1], [2], [4], [8], [3], [6], [12], [11], [9], [5], [10], [7]\}.$$

So 2 is a primitive root modulo 13. Finding primitive roots modulo a given prime p is very difficult. There seems to be no other way than trying out elements in $\mathbb{F}_p^*$ and seeing whether they generate $\mathbb{F}_p^*$. In this sense the proof of Theorem 4.5.3 is abstract and nice. It gives the comfort of knowing of the existence of a generator by appealing to properties of cyclotomic polynomials. But it leaves no clue as how to find the former. The difficulty of this problem is probably related to the difficulty of computing φ for large integers. Suppose that we pick a random element $a \in \mathbb{F}_p^*$. By Theorem 4.5.3, the probability that a will be a primitive root is

$$\frac{\varphi(p-1)}{p-1},$$

since there are $\varphi(p - 1)$ generators in a cyclic group of order $p - 1$ by Proposition 2.7.4(iii). This number depends heavily on the prime p. Using the Dirichlet theorem on primes in arithmetic progressions one may show that there are primes for which this probability is arbitrarily small ([17], Proposition II.1.3).

4.5.1 Decimal expansions and primitive roots

Here is a famous open problem called the Artin conjecture (after E. Artin (1898–1962)). Given an integer $a > 1$ that is not a square, is a a primitive root for infinitely many prime numbers p? For $a = 10$ this was proved by Gauss. It amounts to showing that there are infinitely many primes p such that the period of the decimal expansion of $1/p$ has length $p - 1$. Let us give two examples of this. If $p = 7$ then

$$1/p = 0.142857142857\ldots.$$

Here the period length is 6. If $p = 17$ then

$$1/p = 0.05882352941176470588\ldots.$$

Here the period length is 16. In general the period length of the fraction $1/p$ is of order $[10]$ in $\mathbb{F}_p^*$ (see Exercise 4.23). So you can use floating point arithmetic to determine the order of $[10]$ in $\mathbb{F}_p^*$ (pocket calculators have limited display size, but a small PC easily handles "infinite" precision floating point numbers).

4.5.2 Primitive roots and public key cryptography

Let us briefly illustrate how the cyclic group $G = \mathbb{F}_p^*$ can be used to construct a public key cryptosystem called the ElGamal cryptosystem. We know from Chapter 1 how to find a large prime number p. Assume now that the involved parties have agreed on sharing a common generator g for G.

The secret deciphering key for A is a number $0 < a < p - 1$. The public key for A is then g^a. To send a message $P \in G$ to A we first generate a random integer k and then send

$$(g^k, Pg^{ak})$$

to A. Now A receives a pair (x, y) where $x, y \in G$. Since A knows a, A can retrieve the original message P by computing $x^{-a}y$. All these operations can be done quite effectively using the repeated squaring algorithm and the extended Euclidean algorithm from Chapter 1.

The security of the cryptosystem relies on the observation that it is difficult to compute a given g^a in G. This problem is known as the discrete logarithm problem in the group G, since a can be viewed as the "discrete" logarithm $\log_g(g^a)$ in the finite group G.

The above cryptosystem makes sense for any cyclic group G. One of the most promising avenues for modern cryptosystems is taking G as a (large) cyclic subgroup of an elliptic curve over a finite field.

4.5.3 Yet another application of cyclotomic polynomials

Using Gaussian integers we proved in Theorem 3.5.20 that there are infinitely many prime numbers $\equiv 1 \pmod 4$. Using cyclotomic polynomials we can generalize this result.

Theorem 4.5.4 *There are infinitely many prime numbers $\equiv 1 \pmod n$ for a natural number $n \geq 2$.*

Proof. It suffices to prove that there exists a prime number $\equiv 1 \pmod n$ for every $n \geq 2$ (why?). Let n be given. We must find a prime number $p \equiv 1 \pmod n$. Since $n \geq 2$ we get $|\Phi_n(n)| > 1$ from Definition 4.4.2. So we may find a prime number p dividing $\Phi_n(n)$. Now Φ_n has a constant term $= \pm 1$ since $|\Phi_n(0)| = 1$ and $\Phi_n(0) \in \mathbb{Z}$. This implies that $p \nmid n$. Therefore $[n]$ is not a multiple root of $X^n - 1 \in \mathbb{F}_p[X]$ by Lemma 4.3.8. Since $\Phi_n([n]) = 0$ in $\mathbb{F}_p$, this implies by Lemma 4.5.2 that $\text{ord}([n]) = n$ in $\mathbb{F}_p^*$, and therefore that n divides $|\mathbb{F}_p^*| = p - 1$ by Proposition 2.6.3(i). This proves that $p \equiv 1 \pmod n$. $\square$

4.6 Ideals in polynomial rings

When is a polynomial ring a Euclidean domain, a principal ideal domain, a unique factorization domain? What are the units? The irreducible elements (polynomials)? How do these concepts relate to roots of polynomials?

In the case of an arbitrary ring these questions cannot be answered easily. There is one crucial result, once again due to Gauss: if R is a unique factorization domain then $R[X]$ is a unique factorization domain. We will not prove this. Our point of departure will be the case where R is a field, which will be denoted F.

Proposition 4.6.1 *The polynomial ring $F[X]$ is a Euclidean domain, a principal ideal domain and a unique factorization domain.*

Proof. We will prove that the degree function deg : $F[X] \setminus \{0\} \to \mathbb{N}$ is a Euclidean function on $F[X]$ (see subsection 3.5.4). Let $d \in F[X] \setminus \{0\}$. Then there exists $q, r \in F[X]$ such that

$$f = qd + r,$$

where either $r = 0$ or $\deg(r) < \deg(d)$. This is the content of Corollary 4.2.5, and it follows that deg is a Euclidean function on $F[X]$. Thus $F[X]$ is a Euclidean domain. This implies by Proposition 3.5.9 that $F[X]$ is a principal ideal domain. We obtain that $F[X]$ is a unique factorization domain by Theorem 3.5.7. $\qquad\square$

Having proved that the degree function on $F[X]$ is a Euclidean function $F[X] \setminus \{0\} \to \mathbb{N}$, we may now use the Euclidean algorithm (as in subsection 3.5.4). This is illustrated in the following example.

Example 4.6.2 Let us use the Euclidean algorithm to find a greatest common divisor of $X^5 + X + 1$ and $X^4 + X^3 + X + 1$ in $\mathbb{F}_2[X]$. Using the division algorithm for polynomials we get

$$X^5 + X + 1 = (X + 1)(X^4 + X^3 + X + 1) + X^3 + X^2 + X$$

and

$$X^4 + X^3 + X + 1 = X(X^3 + X^2 + X) + X^2 + X + 1$$
$$X^3 + X^2 + X = X(X^2 + X + 1).$$

This shows that $X^2 + X + 1$ is a greatest common divisor of $X^5 + X + 1$ and $X^4 + X^3 + X + 1$ in $\mathbb{F}_2[X]$.

Now we move on to state some useful facts about the unique factorization domain $F[X]$. Notice that the concepts of units, irreducible elements etc. from Chapter 3 make perfectly sense for $F[X]$. Irreducible elements in $F[X]$ are called irreducible polynomials. Before embarking upon the next result, let us notice how the degree function comes into play. If $f \in F[X]$ and $f = f_1 f_2$ then

$$\deg(f) = \deg(f_1) + \deg(f_2).$$

If $f = f_1 f_2$ is an honest factorization of f, i.e. if neither f_1 nor f_2 is a constant then $0 < \deg(f_1), \deg(f_2) < \deg(f)$. Polynomials that are units are non-zero constants (degree zero). So if f is not irreducible there is a factorization $f = f_1 f_2$ such that

$$0 < \deg(f_1), \deg(f_2) < \deg(f).$$

This gives us a nice way of deducing that some polynomials are irreducible even if they do not have any roots.

Proposition 4.6.3   *Let $f \in F[X]$. Then we have the following.*

(i) *The ideal $\langle f \rangle$ is a maximal ideal if and only if f is irreducible. In this case the quotient ring*

$$F[X]/\langle f \rangle$$

   is a field.

(ii) *If $f \neq 0$ then f is a unit if and only if $\deg(f) = 0$.*

(iii) *If $\deg(f) = 1$ then f is irreducible.*

(iv) *If f is irreducible and $\deg(f) > 1$ then f does not have any roots.*

(v) *If $\deg(f) = 2$ or $\deg(f) = 3$ then f is irreducible if and only if it has no roots.*

Proof.   (i) This is a consequence of Proposition 3.5.6 and the fact that $F[X]$ is a principal ideal domain. If $\langle f \rangle$ is a maximal ideal then $F[X]/\langle f \rangle$ is a field by Proposition 3.2.7.

   (ii) Non-zero constants (polynomials of degree 0) are units, since F is a field. This follows from Proposition 4.2.3.

   (iii) If f is not an irreducible polynomial then there is a factorization $f = f_1 f_2$ where $0 < \deg(f_1), \deg(f_2) < \deg(f)$. In particular, if $\deg(f) = 1$ then f has to be irreducible.

(iv) If $\alpha \in k$ is a root of f then $f(\alpha) = 0$ and $f = (X - \alpha)h$ for some $h \in k[X]$, by Corollary 4.3.2. This gives $\deg(f) = 1 + \deg(h)$. Since $\deg(f) > 1$, f cannot be irreducible if it has a root.

(v) If $\deg(f) = 2$ or 3 and f is reducible then there is a factorization $f = f_1 f_2$, where either $\deg(f_1) = 1$ or $\deg(f_2) = 1$, since $\deg(f_1) + \deg(f_2) = \deg(f) = 2$ or 3. This shows that f is reducible if and only if a polynomial of degree 1 divides f. This is equivalent to f having a root. □

Example 4.6.4 Consider the polynomial

$$f = X^3 + X + 1 \in \mathbb{F}_5[X].$$

The following table shows that f does not have any roots:

x	0	1	2	3	4
$f(x)$	1	3	1	1	4

.

So we may conclude from Proposition 4.6.3(v) that f is an irreducible polynomial in $\mathbb{F}_5[X]$. What about the polynomial $g = X^4 + X^2 + 1 \in \mathbb{F}_2[X]$? Clearly g does not have any roots. Can we conclude from Proposition 4.6.3 that g is irreducible?

Remark 4.6.5 Cyclotomic polynomials are irreducible as polynomials in $\mathbb{Q}[X]$. This is classical result due to Gauss. The proof consists of a number of clever steps (see Exercise 4.45 (HOF)). What about cyclotomic polynomials when viewed as polynomials in $\mathbb{F}_p[X]$? The cyclotomic polynomial $\Phi_8 = X^4 + 1$ is an example of a polynomial that is reducible in $\mathbb{F}_p[X]$ for all prime numbers p (see Exercise 4.13). In fact one can prove that Φ_n is irreducible in $\mathbb{F}_p[X]$ if and only if $[p]$ generates the group $(\mathbb{Z}/n\mathbb{Z})^*$ (see Exercise 4.43).

A central example is the polynomial $X^2 + 1 \in \mathbb{R}[X]$. This polynomial does not have any roots in $\mathbb{R}$ (since no real number squared equals -1). So by Proposition 4.6.3(v) it follows that $X^2 + 1$ is an irreducible polynomial in $\mathbb{R}[X]$. Also, it follows from Proposition 4.6.3(i) that

$$\mathbb{R}[X]/\langle X^2 + 1 \rangle$$

is a field. In fact it is a very well known field. The next section will reveal the details.

4.6.1 Polynomial rings modulo ideals

The following situation is extremely common: there is a field F and a polynomial $f \in F[X]$ with no roots in F along with an extension field $E \supset F$ such that there exists $\alpha \in E$ with $f(\alpha) = 0$ (think of $F = \mathbb{R}$, $f = X^2 + 1$ and $E = \mathbb{C}$). For later use it is important to notice that $f(\alpha)$ makes sense even though $\alpha \in E$ and $f \in F[X]$. The simple reason is that $F[X] \subseteq E[X]$ as a subring. The purpose of this subsection is to describe an algebraic tool for obtaining an extension field E and a root $\alpha \in E$ given F and $f \in F[X]$. The idea is very clear but hidden in a few technicalities. Let us begin with a detailed example.

Example 4.6.6 We know that

$$F = \mathbb{R}[X]/\langle X^2 + 1 \rangle$$

is a field, since $X^2 + 1$ is an irreducible polynomial in $\mathbb{R}[X]$. How do we describe this field? At this point it is just an abstract quotient ring consisting of cosets of the ideal $\langle X^2 + 1 \rangle$. If we make a few identifications then things become much clearer. By definition

$$F = \{[f] \mid f \in \mathbb{R}[X]\},$$

where $[f]$ is the coset $f + \langle X^2 + 1 \rangle$. Dividing f by $X^2 + 1$ we get

$$f = q(X^2 + 1) + aX + b,$$

where $a, b \in \mathbb{R}$. This is a consequence of Proposition 4.2.4 and a substantial simplification of $[f]$, since

$$[f] = [q(X^2 + 1) + aX + b] = [aX + b]$$

because $q(X^2 + 1) \in \langle X^2 + 1 \rangle$. So we may write

$$F = \{[aX + b] \mid a, b \in \mathbb{R}\}.$$

An added bonus is that the elements in F are uniquely given as $[aX + b]$, where $a, b \in \mathbb{R}$. Suppose that $[aX + b] = [cX + d]$, where $c, d \in \mathbb{R}$. Then $(aX + b) - (cX + d) = (a - c)X + (b - d) \in \langle X^2 + 1 \rangle$. Thus

$$(a - c)X + (b - d) = q(X^2 + 1)$$

for some $q \in \mathbb{R}[X]$. Here Proposition 4.2.2 gives that $a = c$ and $b = d$.

The next step is to realize that $\mathbb{R}$ is a subring of F. This is easy: instead of writing $[r]$ we simply write r when $r \in \mathbb{R}$ is a constant polynomial. This is allowed, since $[r_1] = [r_2]$ if and only $r_1 = r_2$, when $r_1, r_2 \subset \mathbb{R}$. So

$$F = \{a + b[X] \mid a, b \in \mathbb{R}\}.$$

We now have a satisfactory description of the elements in F. Addition and multiplication in F are given using addition and multiplication in a quotient ring: thus $[g_1] + [g_2] = [g_1 + g_2]$ and $[g_1][g_2] = [g_1 g_2]$. In our notation addition is given by

$$(a + b[X]) + (c + d[X]) = (a + c) + (b + d)[X].$$

To do multiplication we obtain initially

$$(a + b[X])(c + d[X]) = ac + (ad + bc)[X] + bd[X^2].$$

But $[X^2]$ does not fit our description of elements in F as given by $x + y[X]$ with $x, y \in \mathbb{R}$. Fortunately this is easy to repair since $[X^2] = [-1] = -1 \in F$. With this in mind we get

$$(a + b[X])(c + d[X]) = (ac - bd) + (ad + bc)[X].$$

Through this algebraic process we have shown that F is the field $\mathbb{C}$ of complex numbers. The role of $i = \sqrt{-1}$ is played by $[X] \in F$ as $[X]^2 = [X^2] = -1$.

We now return to the general case of coefficients in a ring R. We know that R is a natural subring (consisting of the constant polynomials) of $R[X]$. Let I be an ideal in $R[X]$ with $R \cap I = \langle 0 \rangle$ (0 is the only constant polynomial in I). If $r_1, r_2 \in R$ and $[r_1] = [r_2]$ in $R[X]/I$ then $r_1 - r_2 \in R \cap I$. Therefore $r_1 = r_2$. So if $R \cap I = \langle 0 \rangle$ then we may use the notation r to denote the element $[r]$ in $R[X]/I$ (where $r \in R$). The details of Example 4.6.6 cover all the steps in the proof of the following proposition.

Proposition 4.6.7 *Let R be a ring and*

$$f = X^n + a_{n-1}X^{n-1} + \cdots + a_1 X + a_0 \in R[X]$$

a monic polynomial of positive degree n. Then $R \cap \langle f \rangle = \langle 0 \rangle$. The elements $[g] = g + \langle f \rangle$ in the quotient ring $R[X]/\langle f \rangle$ can be expressed uniquely as polynomials of degree $< n$

$$b_0 + b_1\alpha + \cdots + b_{n-1}\alpha^{n-1},$$

where $b_0, \ldots, b_{n-1} \in R$ and $\alpha = [X]$. In $R[X]/\langle f \rangle$ we have the identity

$$\alpha^n = -a_{n-1}\alpha^{n-1} - \cdots - a_1\alpha - a_0.$$

Proof. Suppose that $r \in R \cap \langle f \rangle$. Then there exists $q \in R[X]$ such that $r = qf$. If $q \neq 0$ then $\deg(q) + \deg(f) = \deg(q) + n > 0$ by Proposition 4.2.2. This contradicts that r is a constant. So $q = 0$ and $R \cap \langle f \rangle = \langle 0 \rangle$.

Suppose that $[g] \in R[X]/\langle f \rangle$. Write $g = qf + r$, where $r = 0$ or $r \neq 0$ and $\deg(r) < n = \deg(f)$ after dividing by f. So $[g] = [qf + r] = [qf] + [r] = [r]$. Suppose that $r_1, r_2 \in R[X] \setminus \{0\}$, that $\deg(r_1)$, $\deg(r_2) < n$ and that $[r_1] = [r_2]$. Then there exists $q \in R[X]$ such that $r_1 - r_2 = qf$. By the same reasoning (using Proposition 4.2.2) as above we see that $r_1 = r_2$. So every non-zero element in the quotient ring can described uniquely as $[g]$, where $\deg(g) < n$ and $g \in R[X] \setminus \{0\}$. Writing this out we obtain

$$[g] = [b_0 + b_1 X + \cdots + b_{n-1} X^{n-1}] = b_0 + b_1 \alpha + \cdots + b_{n-1} \alpha^{n-1},$$

where $\alpha = [X]$ and $b_0, \ldots, b_{n-1} \in R$. Since

$$\begin{aligned}
[f] &= [X^n + a_{n-1} X^{n-1} + \cdots + a_1 X + a_0] \\
&= [X^n] + a_{n-1}[X^{n-1}] + \cdots + a_1[X] + a_0 \\
&= \alpha^n + a_{n-1} \alpha^{n-1} + \cdots + a_1 \alpha + a_0 = 0,
\end{aligned}$$

we get the desired identity for α^n in $R[X]/\langle f \rangle$.   □

Notice that R is a natural subring of $R[X]/\langle f \rangle$ above. The natural ring homomorphism $\varphi : R \to R[X]/\langle f \rangle$ given by $\varphi(r) = [r]$ is injective.

Remark 4.6.8 If F is a field and $f \in F[X]$ an irreducible polynomial then $\langle f \rangle$ is a maximal ideal and $F[X]/\langle f \rangle$ becomes a field extension E of F. Now $\alpha = [X] \in E$, and this actually is a root of $f \in F[X] \subseteq E[X]$ since $f(\alpha) = 0$ by the identity for α^n in Proposition 4.6.7. This is the algebraic way of using an irreducible polynomial to construct a bigger field where it has a root.

Let us illustrate how the identity for α^n in Proposition 4.6.7 completely determines multiplication in the quotient ring.

Example 4.6.9 Let $f = X^2 + X + 1 \in \mathbb{F}_2[X]$. Then f is an irreducible polynomial since it has no roots (Proposition 4.6.3). This means that $\langle f \rangle$ is a maximal ideal and that the quotient ring $F = \mathbb{F}_2[X]/\langle f \rangle$ is a field. Now, by Proposition 4.6.7, $F = \{a_0 + a_1 \alpha \mid a_0, a_1 \in \mathbb{F}_2\}$, where $\alpha = [X]$ and the rule $\alpha^2 = -1 - \alpha$ determines the multiplication. Multiplying $a + b\alpha$ by $c + d\alpha$ we get

$$\begin{aligned}
ac + (ad + bc)\alpha + bd\alpha^2 &= ac + (ad + bc)\alpha + bd(-1 - \alpha) \\
&= (ac - bd) + (ad + bc - bd)\alpha.
\end{aligned}$$

Notice that F is an extension field of $\mathbb{F}_2$ with four elements.

Having proved Proposition 4.6.7 we have the tools for proving one of the true highlights of number theory.

4.7 Theorema Aureum: the law of quadratic reciprocity

We now show how a specific quotient of a polynomial ring gives a beautiful proof of the law of quadratic reciprocity (see Section 1.11). Gauss called the law of quadratic reciprocity Theorema Aureum, the golden theorem. He gave six proofs (see [14], Chapter 5) of this theorem during his lifetime. In 1921 there were 56 known proofs of quadratic reciprocity. Today there could be well over a hundred. Let us recall the statement of quadratic reciprocity. We are given two odd prime numbers p and q. Then the Legendre symbols (Definition 1.11.1) of p and q are related through the breathtaking identity

$$\left(\frac{p}{q}\right)\left(\frac{q}{p}\right) = (-1)^{(p-1)(q-1)/4}.$$

We will work in the ring

$$R = \mathbb{F}_p[X]/\langle 1 + X + \cdots + X^{q-1}\rangle.$$

Recall from Proposition 4.6.7 that an element in R can be written uniquely in terms of $\zeta = [X]$ as

$$a_0 + a_1\zeta + \cdots + a_{q-2}\zeta^{q-2},$$

where $a_0, \ldots, a_{q-2} \in \mathbb{F}_p$.

Lemma 4.7.1   *The element ζ is a primitive qth root of unity in R. Let $\xi = \zeta^l$ where $q \nmid l$. Then*

$$1 + \xi + \cdots + \xi^{q-1} = 0$$

in R.

Proof.   It follows from Proposition 4.6.7 that $\zeta, \ldots, \zeta^{q-2} \neq 1$ and

$$\zeta^{q-1} = -1 - \zeta - \cdots - \zeta^{q-2} \neq 1.$$

A small computation now shows that $\zeta^q = \zeta\zeta^{q-1} = 1$. This proves that ζ is a primitive qth root of unity. If $q \nmid l$ then $\gcd(q, l) = 1$. Therefore $\{1, \zeta, \ldots, \zeta^{q-1}\} = \{1, \xi, \ldots, \xi^{q-1}\}$. It follows that

$$1 + \xi + \cdots + \xi^{q-1} = 1 + \zeta + \cdots + \zeta^{q-1} = 0. \qquad \square$$

Now consider the so-called Gauss sum

$$G = \sum_{j=0}^{q-1} \left(\frac{j}{q}\right) \zeta^j$$

in R. The individual terms satisfy

$$\left(\frac{j}{q}\right) \zeta^j = \left(\frac{j+qm}{q}\right) \zeta^{j+qm}$$

for every $m \in \mathbb{Z}$. This is used heavily in the proof of the following important lemma.

Lemma 4.7.2   *The Gauss sum $G \in R$ satisfies the following.*

(i) $G^2 = (-1)^{(q-1)/2} q$.

(ii) *G is an invertible element in the ring R if $p \neq q$.*

Proof.   The invertibility of G follows from (i) as $q \in \mathbb{F}_p \subset R$ is invertible in R since it is invertible in $\mathbb{F}_p$ if $p \neq q$. The proof of (i) is fairly straightforward, but it contains some clever rewritings:

$$G^2 = \left(\sum_{j=0}^{q-1} \left(\frac{j}{q}\right) \zeta^j\right) \left(\sum_{j=0}^{q-1} \left(\frac{j}{q}\right) \zeta^j\right)$$

$$= \left(\sum_{i=1}^{q-1} \left(\frac{i}{q}\right) \zeta^i\right) \left(\sum_{j=1}^{q-1} \left(\frac{-j}{q}\right) \zeta^{-j}\right),$$

since

$$\left(\frac{-j}{q}\right) \zeta^{-j} = \left(\frac{q-j}{q}\right) \zeta^{q-j}.$$

We continue by rewriting the last sum in the expression for G^2:

$$\sum_{i,j=1}^{q-1} \left(\frac{i}{q}\right) \left(\frac{-j}{q}\right) \zeta^{i-j} = \left(\frac{-1}{q}\right) \sum_{i,j=1}^{q-1} \left(\frac{ij}{q}\right) \zeta^{i-j}$$

$$= (-1)^{(q-1)/2} \sum_{i,j=1}^{q-1} \left(\frac{i^2 j}{q}\right) \zeta^{i(1-j)}.$$

Here we have used the formula

$$\left(\frac{-1}{q}\right) = (-1)^{(q-1)/2}$$

from Proposition 1.11.6. We have also replaced j with ij in the terms of the sum. We may do this because if j runs through $1, \ldots, q-1$ then the remainders of ij modulo q run through $1, \ldots, q-1$ (though not in the same order). Since

$$\left(\frac{i^2}{q}\right) = 1,$$

we end up with the expression

$$(-1)^{(q-1)/2} \sum_{j=1}^{q-1} \left(\frac{j}{q}\right) \sum_{i=1}^{q-1} \zeta^{i(1-j)},$$

which is equal to

$$(-1)^{(q-1)/2} \sum_{j=1}^{q-1} \left(\frac{j}{q}\right) \sum_{i=0}^{q-1} \zeta^{i(1-j)}$$

since

$$\sum_{j=1}^{q-1} \left(\frac{j}{q}\right) = 0$$

by Proposition 1.11.3. By Lemma 4.7.1 it follows that

$$\sum_{i=0}^{q-1} \zeta^{i(1-j)}$$

is non-zero precisely if $j = 1$. In this case it is equal to q, proving the formula for G^2.   □

Raising G to the pth power in R we get the formula

$$G^p = (G^2)^{(p-1)/2}G = (-1)^{(p-1)(q-1)/4}q^{(p-1)/2}G$$
$$= (-1)^{(p-1)(q-1)/4} \left(\frac{q}{p}\right) G \qquad (4.2)$$

by Lemma 4.7.2 and Theorem 1.11.4. Computing the left hand side from the definition and using Freshman's Dream (Theorem 3.3.9) in the ring R we get

$$G^p = \sum_{j=0}^{q-1} \left(\frac{j}{q}\right) \zeta^{pj} = \sum_{j=0}^{q-1} \left(\frac{p}{q}\right)\left(\frac{pj}{q}\right) \zeta^{pj}$$
$$= \left(\frac{p}{q}\right) G.$$

Comparing this expression with (4.2) and using that G is invertible in R (Lemma 4.7.2), we obtain the law of quadratic reciprocity,

$$\left(\frac{p}{q}\right) = (-1)^{(p-1)(q-1)/4}\left(\frac{q}{p}\right).$$

4.8 Finite fields

Finite fields are among the most beautiful objects in algebra. We already know the finite fields $\mathbb{F}_p$, where p is a prime number. But Example 4.6.9 indicated that this is not the whole story (there we constructed a field with $4 = 2^2$ elements). In this section we prove that there exists a unique (up to ring isomorphism) finite field with p^n elements, where p is a prime number and n a natural number. We start out with a lemma showing that a finite field looks exactly like the extension field we encountered in Example 4.6.9.

Lemma 4.8.1   *Let F be a finite field. Then $|F| = p^n$, where p is a prime number, $n \geq 1$ and there exists an irreducible polynomial $f \in \mathbb{F}_p[X]$ of degree n such that*

$$F \cong \mathbb{F}_p[X]/\langle f \rangle.$$

Proof.   Consider the unique ring homomorphism $\kappa : \mathbb{Z} \to F$. Since F is finite, κ is not injective. This implies that the characteristic of F is a prime number p, by Proposition 3.3.7. We may view $\mathbb{F}_p$ as a subring of F by Lemma 3.3.5. As F is finite we obtain from Theorem 4.5.3 that F^* is a cyclic group. Let $\gamma \in F^*$ be a generator for F^*. Thus every element in F is either 0 or a power γ^n of γ. Since $\varphi_\gamma(X^n) = \gamma^n$, the ring homomorphism $\varphi_\gamma : F[X] \to F$ is surjective. More than this is true, though. In fact by restricting φ_γ to $\mathbb{F}_p[X] \subseteq F[X]$ we get a surjective ($X^n \in \mathbb{F}_p[X]$) ring homomorphism

$$\varphi : \mathbb{F}_p[X] \to F.$$

The kernel $\mathrm{Ker}\,\varphi$ of φ is a principal ideal $\langle f \rangle \subseteq \mathbb{F}_p[X]$ by Proposition 4.6.1. By Proposition 3.3.2 we get

$$\mathbb{F}_p[X]/\langle f \rangle \cong F,$$

so that $\langle f \rangle$ is a maximal ideal by Proposition 3.2.7. This implies that f is an irreducible polynomial by Proposition 4.6.3(i). By Proposition 4.6.7, $|F| = p^n$, where $n = \deg(f)$. This proves the lemma. $\square$

The main result is the existence and uniqueness up to isomorphism of the finite fields alluded to in Lemma 4.8.1. Below we state the theorem. The main ingredients in the proof will occupy subsections 4.8.1 and 4.8.2.

Theorem 4.8.2 *There exists a unique finite field with p^n elements, where p is a prime number and $n \geq 1$. More precisely, we have the following.*

(i) *There exists an irreducible polynomial in $\mathbb{F}_p[X]$ of degree n.*
(ii) *Suppose that F and F' are finite fields with p^n elements. Then there exists a ring isomorphism $F \overset{\sim}{\to} F'$.*

Proof. Suppose that f is an irreducible polynomial in $\mathbb{F}_p[X]$ of degree n. Then $\langle f \rangle$ is a maximal ideal by Proposition 4.6.3(i). Therefore $\mathbb{F}_p[X]/\langle f \rangle$ is a field. It has p^n elements by Proposition 4.6.7. The proof of (i) is a surprising application of cyclotomic polynomials and will be described in subsection 4.8.1. The proof of (ii) is described in subsection 4.8.2. □

Before entering the finer details of the proof of Theorem 4.8.2 we need a crucial lemma involving only natural numbers.

Lemma 4.8.3 *Let τ, d and n be natural numbers, where $\tau > 1$. Then $\tau^d - 1$ divides $\tau^n - 1$ if and only if d divides n.*

Proof. We may assume that $d \geq 1$. By Theorem 1.2.1 we write $n = qd + r$, where $0 \leq r < d$. Then

$$\frac{\tau^n - 1}{\tau^d - 1} = \frac{(\tau^d)^q \tau^r - 1}{\tau^d - 1}$$
$$= \tau^r \frac{(\tau^d)^q - 1}{\tau^d - 1} + \frac{\tau^r - 1}{\tau^d - 1}$$
$$= \tau^r (1 + \tau^d + \cdots + (\tau^d)^{q-1}) + \frac{\tau^r - 1}{\tau^d - 1}.$$

As $0 \leq \tau^r - 1 < \tau^d - 1$ this proves the claim. □

Remark 4.8.4 Theorem 4.8.2 says that there exists a unique field F with p^n elements up to isomorphism. We denote F by $\mathbb{F}_{p^n}$. Informally one may say that there is only one way to multiply in a finite field with p^n elements.

4.8.1 Existence of finite fields

We know that there are infinitely many irreducible polynomials in $\mathbb{F}_p[X]$ (see Exercise 4.7), but this does not guarantee that we may find one of each degree. This is where cyclotomic polynomials enter. If we view them as polynomials in $\mathbb{F}_p[X]$ they have very interesting properties.

Theorem 4.8.5 *There exists an irreducible polynomial in* $\mathbb{F}_p[X]$ *of degree* $n \geq 1$. *More precisely, if* f *is an irreducible polynomial dividing* Φ_{p^n-1} *in* $\mathbb{F}_p[X]$ *then* $\deg(f) = n$.

Proof. Let $d = \deg(f)$. Then $L = \mathbb{F}_p[X]/\langle f \rangle$ is a field with p^d elements and $\alpha = [X]$ is a root of $f \in \mathbb{F}_p[X] \subseteq L[X]$ by Remark 4.6.8. Since $gf = \Phi_{p^n-1}$ for $g \in \mathbb{F}_p[X]$ we get $\Phi_{p^n-1}(\alpha) = g(\alpha)f(\alpha) = 0$. The derivative of $X^{p^n-1} - 1$ is

$$D(X^{p^n-1} - 1) = (p^n - 1)X^{p^n-2} = -X^{p^n-2}.$$

This shows by Lemma 4.3.8 that α is not a multiple root of $X^{p^n-1} - 1$ and therefore that α is a primitive $(p^n - 1)$th root of unity in L by Lemma 4.5.2. Now, $\alpha^{p^d-1} = 1$ shows that $p^n - 1 \mid p^d - 1$ by Proposition 2.6.3.

Let $R = \{\xi \in L \mid \xi^{p^n} = \xi\}$. This is a subring of L by Theorem 3.3.9. Since $\alpha^{p^n-1} = 1$ we must have $\alpha \in R$. But since $L = \{a_0 + a_1\alpha + \cdots + a_{d-1}\alpha^{d-1} \mid a_i \in \mathbb{F}_p\}$ by Proposition 4.6.7 it follows that $R = L$ (R contains $1, \alpha, \alpha^2, \dots$ and is a subring). By Theorem 4.5.3 there exists a primitive $(p^d - 1)$th root of unity ζ in L. Since $\zeta \in R$ we obtain $\zeta^{p^n-1} = 1$. Proposition 2.6.3(iii) gives $p^d - 1 \mid p^n - 1$. Therefore $p^d - 1 = p^n - 1$. This shows that $d = n$. $\square$

Remark 4.8.6 Theorem 4.8.5 says that if

$$\Phi_{p^n-1} = f_1 \cdots f_r$$

is an irreducible factorization of Φ_{p^n-1} in $\mathbb{F}_p[X]$ then $\deg(f_i) = n$. In particular, $n \mid \varphi(p^n - 1)$.

4.8.2 Uniqueness of finite fields

Suppose that F and F' are finite fields with p^n elements. Then $F \cong \mathbb{F}_p[X]/\langle f \rangle$ for a suitable irreducible polynomial $f \in \mathbb{F}_p[X]$ of degree n, by Lemma 4.8.1.

Furthermore $f(\alpha) = 0$, where $\alpha = [X] \in F$ by Remark 4.6.8. Notice that $I = \{g \in \mathbb{F}_p[X] \mid g(\alpha) = 0\} \subsetneq \mathbb{F}_p[X]$ is an ideal in $\mathbb{F}_p[X]$. Now $f \in I$ and therefore $\langle f \rangle \subseteq I$. But since $\langle f \rangle$ is a maximal ideal we must have $I = \langle f \rangle$. Because F^* is a finite group with $p^n - 1$ elements, we obtain $\xi^{p^n-1} = 1$ for every $\xi \in F^*$ by Propositon 2.6.3(ii). This implies that $X^{p^n} - X \in I$ and therefore that $f \mid X^{p^n} - X$ in $\mathbb{F}_p[X]$. In $F'[X]$ we have the factorization

$$X^{p^n} - X = \prod_{\alpha \in F'} (X - \alpha),$$

since every $\beta \in F'$ satisfies $\beta^{p^n} = \beta$ by Proposition 2.6.3(ii). Therefore $f \in \mathbb{F}_p[X] \subseteq F'[X]$ must have a root α' in F' since it divides $X^{p^n} - X$. Now look at

$$\varphi_{\alpha'} : \mathbb{F}_p[X] \to F'.$$

Obviously $\langle f \rangle \subseteq \mathrm{Ker}\,\varphi_{\alpha'}$, but since $\mathrm{Ker}\,(\varphi_{\alpha'})$ is a proper ideal and $\langle f \rangle$ is a maximal ideal, we must have $\langle f \rangle = \mathrm{Ker}\,(\varphi_{\alpha'})$. Therefore we get an injective ring homomorphism

$$\mathbb{F}_p[X]/\langle f \rangle \to F'.$$

But since F' has p^n elements, this must be a bijection and thereby an isomorphism (of rings). We have proved that two finite fields F and F' with the same number of elements are isomorphic.

4.8.3 A beautiful identity

We already know that

$$X^{p^n} - X = X(X^{p^n-1} - 1) = X \prod_{d \mid p^n - 1} \Phi_d$$

in $\mathbb{F}_p[X]$. By Theorem 4.8.5, $X^{p^n} - X$ is divisible by an irreducible polynomial of degree n. This is not the entire story. We will compute the complete irreducible factorization of $X^{p^n} - X$ in $\mathbb{F}_p[X]$. Let us compute this factorization in some special cases.

Example 4.8.7 In $\mathbb{F}_2[X]$ we have

$$X^{2^2} - X = X^4 - X = X(X + 1)(X^2 + X + 1).$$

In $\mathbb{F}_3[X]$ we have

$$X^{3^2} - X = X^9 - X = X(X+1)(X+2)(X^2+1)(X^2+X+2)(X^2+2X+2).$$

The general result is the following surprising theorem.

Theorem 4.8.8   *The polynomial* $X^{p^n} - X \in \mathbb{F}_p[X]$ *is the product*

$$X^{p^n} - X = f_1 \cdots f_r$$

of the monic irreducible polynomials $f_1, \ldots, f_r$ *in* $\mathbb{F}_p[X]$ *of degree* d, *where* $1 \leq d \leq n$ *and* $d \mid n$.

Proof.   Let $f \in \mathbb{F}_p[X]$ be a monic irreducible polynomial of degree d. Then $L = \mathbb{F}_p[X]/\langle f \rangle$ is a field (by Proposition 4.6.3(i)) with p^d elements (by Proposition 4.6.7). Let $\alpha = [X]$ in L. Then $\alpha^{p^d} = \alpha$ by Proposition 2.6.3(ii). If $d \mid n$ then $\alpha^{p^n} = \alpha$. This is seen by raising both sides of $\alpha^{p^d} = \alpha$ to the p^dth power $q - 1$ times, where $n = qd$. The identity $\alpha^{p^n} = \alpha$ in L means that $X^{p^n} - X \in \langle f \rangle$ or that $f \mid X^{p^n} - X$.

Now assume that f divides $X^{p^n} - X$. We wish to prove that $d \mid n$. Consider the subset

$$R = \{\zeta \in L \mid \zeta^{p^n} = \zeta\}.$$

Then R is a subring of L by Theorem 3.3.9. It contains α, as f divides $X^{p^n} - X$. But since $L = \{a_0 + a_1\alpha + \cdots + a_{d-1}\alpha^{d-1} \mid a_i \in \mathbb{F}_p\}$, by Proposition 4.6.7, it follows that $R = L$ (R contains $1, \alpha, \alpha^2, \ldots$ and is a subring). Let γ be a generator for the cyclic group L^*. The order of γ in the group L^* is $p^d - 1$, and $\gamma^{p^n-1} = 1$ since $\gamma \in R$. This implies by Proposition 2.6.3(iii) that $p^d - 1 \mid p^n - 1$. Finally, we obtain $d \mid n$ by Lemma 4.8.3.

Let $f_1, \ldots, f_r$ denote the monic irreducible polynomials of degree $d \mid n$. We have proved that

$$X^{p^n} - X = f_1^{n_1} \cdots f_r^{n_r},$$

where $n_1, \ldots, n_r \geq 1$. One thing is still missing in the proof of our identity. We need to make sure that the multiplicities $n_1, \ldots, n_r$ are all 1. This can be done by proving that $X^{p^n} - X$ is not divisible by the square of an irreducible polynomial. This follows from Lemma 4.3.8(i), since $D(X^{p^n} - X) = p^n X^{p^n-1} - 1 = -1$. $\square$

We have the following consequence of Theorem 4.8.8.

Corollary 4.8.9 *Let N_d denote the number of monic irreducible polynomials of degree d in $\mathbb{F}_p[X]$. Then*

$$p^n = \sum_{d|n} d N_d.$$

Proof. This follows by applying the degree function deg to both sides of the formula in Theorem 4.8.8.   □

There are p irreducible monic polynomials of degree 1 in $\mathbb{F}_p[X]$. These can be listed as

$$X, \quad X-1, \quad X-2, \quad \ldots, \quad X-(p-1),$$

showing that $N_1 = p$. If q is a prime number then Corollary 4.8.9 implies that

$$p^q = q N_q + N_1 = q N_q + p;$$

thus

$$N_q = \frac{p^q - p}{q}.$$

It follows from Theorem 4.8.8 that in general

$$N_n = \frac{p^n - \sum_{d<n,\, d|n} d N_d}{n}.$$

An explicit formula for N_n is given by

$$N_n = \frac{1}{n} \sum_{d|n} \mu\left(\frac{n}{d}\right) p^d,$$

where μ is the Möbius function (given by $\mu(1) = 1$, $\mu(n) = 0$, if n is divisible by a square > 1, and $\mu(p_1 \cdots p_l) = (-1)^l$, where $p_1, \ldots, p_l$ are distinct primes). Another important consequence of Theorem 4.8.8 is a clever factoring algorithm for polynomials.

Suppose that $g \in \mathbb{F}_p[X]$ is a monic polynomial, $\deg(g) = d$ and

$$g = g_1 \cdots g_d,$$

where g_i denotes the product of the (monic) irreducible polynomials of degree i dividing g. Then it follows from Theorem 4.8.8 that

$$\gcd(g, \, X^{p^i} - X)$$

is the product of g_j for $j \mid i$. A straightforward algorithm for finding $g_1, \ldots, g_d$ is to insert $i = 1, 2, \ldots$ in $\gcd(g, \, X^{p^i} - X)$ and use the Euclidean algorithm to compute the greatest common divisor; it is not clear when this algorithm was first discovered. The remaining problem is how to factor out the irreducible polynomials of the same degree i from g_i. A nice solution to this problem was found by Cantor and Zassenhaus in 1979 (see [16], subsection 4.6.2, or [6], subsection 8.4.4). You should prove Lemma 4.8.10 and gain some computational experience by doing Exercise 4.41.

We will move on to describe a general factoring algorithm for polynomials over $\mathbb{F}_p$ and an easy criterion detecting when a given polynomial is irreducible, using only linear algebra. For an introduction to linear algebra over arbitrary fields please consult Appendix B.

4.9 Berlekamp's algorithm

Let f be a polynomial in $\mathbb{F}_p[X]$. We have a few ways, but they are very limited, of deciding whether f is irreducible. If $\deg(f) = 2, 3$ then Proposition 4.6.3 shows that f is irreducible if and only if f does not have a root. In degree 4 and above there seems to be no way other than brute force for deciding whether f is irreducible. Therefore it is quite surprising to find that there is an easy way of deciding this merely by looking at the matrices of two linear maps.

Since the quotient ring $R = \mathbb{F}_p[X]/\langle f \rangle$ has characteristic p, the Frobenius map $F(v) = v^p$ is a ring homomorphism

$$F : R \to R.$$

This is just Theorem 3.3.9. But here R is not only a ring, it is also a vector space over $\mathbb{F}_p$. Since $\lambda^p = \lambda$ for $\lambda \in \mathbb{F}_p$, $F : R \to R$ is in fact a linear map of $\mathbb{F}_p$ vector spaces. A simple example will illustrate how linear algebra comes into play.

Example 4.9.1 Let $f = X^5 + X + 1 \in \mathbb{F}_2[X]$. Then $R = \mathbb{F}_2[X]/\langle f \rangle$ is a vector space over $\mathbb{F}_2$ with basis $1, \alpha, \alpha^2, \alpha^3, \alpha^4$, where $\alpha = [X]$. The element $\alpha^5 \in R$ is expressed in this basis as $\alpha + 1$ by Proposition 4.6.7. The Frobenius map $F(v) = v^2$ is an $\mathbb{F}_2$-linear map $R \to R$. We can compute its 5×5 matrix with respect to the basis $1, \alpha, \alpha^2, \alpha^3, \alpha^4$. Since $F(1) = 1$, $F(\alpha) = \alpha^2$, $F(\alpha^2) = \alpha^4$, $F(\alpha^3) = \alpha^6 = \alpha\alpha^5 = \alpha(\alpha + 1) = \alpha^2 + \alpha$ and $F(\alpha^4) = \alpha^8 = \alpha^3\alpha^5 = \alpha^3(\alpha + 1) = \alpha^4 + \alpha^3$, the matrix of F is

$$\begin{pmatrix} 1 & 0 & 0 & 0 & 0 \\ 0 & 0 & 0 & 1 & 0 \\ 0 & 1 & 0 & 1 & 0 \\ 0 & 0 & 0 & 0 & 1 \\ 0 & 0 & 1 & 0 & 1 \end{pmatrix}.$$

If $\mathrm{Ker}\,(F) \neq 0$ we can find a non-constant polynomial $g \in F_p[X]$ such that $\deg(g) < \deg(f)$ and $[g]^p = [0]$. This means that $f \mid g^p$. If π is an irreducible polynomial dividing f then π divides g. Thus we obtain that $\gcd(f, g)$ is a non-trivial divisor in f $(0 < \deg(\gcd(f, g)) < \deg(f))$.

If $g \in F_p[X]$ is a polynomial such that $0 < \deg(g) < \deg(f)$ and $[g] \in \mathrm{Ker}\,(F - I)$, where I is the identity map, then $[g]^p = [g]$ in R. We have the crucial factorization

$$g^p - g = g(g - 1) \cdots (g - p + 1),$$

since

$$X^p - X = X(X - 1) \cdots (X - p + 1)$$

in $F_p[X]$. Let π be an irreducible polynomial dividing f. Since $f \mid g^p - g$ we obtain that π divides one of $g, g - 1, \ldots, g - p + 1$. Thus one of $\gcd(f, g), \gcd(f, g - 1), \ldots, \gcd(f, g - p + 1)$ is a non-trivial factor of f, since $\deg(g) < \deg(f)$.

Example 4.9.2 The matrix for $F - I$, where F is given in Example 4.9.1, is

$$\begin{pmatrix} 0 & 0 & 0 & 0 & 0 \\ 0 & 1 & 0 & 1 & 0 \\ 0 & 1 & 1 & 1 & 0 \\ 0 & 0 & 0 & 1 & 1 \\ 0 & 0 & 1 & 0 & 0 \end{pmatrix}.$$

We see that

$$\begin{pmatrix} 0 & 0 & 0 & 0 & 0 \\ 0 & 1 & 0 & 1 & 0 \\ 0 & 1 & 1 & 1 & 0 \\ 0 & 0 & 0 & 1 & 1 \\ 0 & 0 & 1 & 0 & 0 \end{pmatrix} \begin{pmatrix} 1 \\ 1 \\ 0 \\ 1 \\ 1 \end{pmatrix} = \begin{pmatrix} 0 \\ 0 \\ 0 \\ 0 \\ 0 \end{pmatrix}.$$

This implies that the polynomial $g = 1 + X + X^3 + X^4$ satisfies $f \mid g^p - g$. By the Euclidean algorithm one obtains (see Example 4.6.2)

$$\gcd(X^5 + X + 1, \; X^4 + X^3 + X + 1) = X^2 + X + 1.$$

This is a non-trivial factor in $X^5 + X + 1$.

The big surprise is that one needs only to look at the $\mathbb{F}_p$ linear maps F and $F - I$ in order to decide whether f is irreducible. The proof of the following theorem is due to B. Iversen.

Theorem 4.9.3 *Let $f \in \mathbb{F}_p[X]$ be a non-constant polynomial and let F denote the Frobenius map*

$$F : R \to R,$$

where $R = \mathbb{F}_p[X]/\langle f \rangle$. Then f is irreducible if and only if $\mathrm{Ker}\,(F) = 0$ and $\mathrm{Ker}\,(F - I) = \mathbb{F}_p$, where I is the identity map $R \to R$.

Proof. We have seen that $\mathrm{Ker}\,(F) = 0$ and $\mathrm{Ker}\,(F - I) = \mathbb{F}_p$ if f is irreducible (if not, we saw how to find a non-trivial factor in f). Assume now that $\mathrm{Ker}\,(F) = 0$ and $\mathrm{Ker}\,(F - I) = \mathbb{F}_p$ and let a be a non-zero element of R. We wish to prove that $1 \in \mathrm{Im}(\varphi)$, where φ is the linear map $\varphi(x) = ax$. This will imply that a is invertible and therefore that R is a field (so that f has to be irreducible). Suppose that $x \in \mathrm{Ker}\,(\varphi) \cap \mathrm{Im}(\varphi)$. Then $x = ay$ for a suitable $y \in R$ and $ax = 0$. This implies that $F(x) = a^p y^p = a^{p-2} y^{p-1} ax = 0$. Therefore $x \in \mathrm{Ker}\,(F)$, so that $x = 0$ and $\mathrm{Ker}\,(\varphi) \cap \mathrm{Im}(\varphi) = 0$. If $v_1, \ldots, v_r$ is a basis of $\mathrm{Ker}\,(\varphi)$ and $w_1, \ldots, w_s$ is a basis of $\mathrm{Im}(\varphi)$ then $v_1, \ldots, v_r, w_1, \ldots, w_s$ is a basis of the subspace $\mathrm{Ker}\,(\varphi) + \mathrm{Im}(\varphi)$ of R. This implies that $\dim_{\mathbb{F}_p} \mathrm{Ker}\,(\varphi) + \mathrm{Im}(\varphi) = \dim_{\mathbb{F}_p} \mathrm{Ker}\,(\varphi) + \dim_{\mathbb{F}_p} \mathrm{Im}(\varphi) = \dim_{\mathbb{F}_p} R$, so that

$$R = \mathrm{Ker}\,(\varphi) + \mathrm{Im}(\varphi).$$

Notice that if $x \in \mathrm{Ker}\,(\varphi)$ then $F(x) \in \mathrm{Ker}\,(\varphi)$ (the same holds for $\mathrm{Im}(\varphi)$). Now write $1 = \alpha + \beta$, where $\alpha \in \mathrm{Ker}\,(\varphi)$ and $\beta \in \mathrm{Im}(\varphi)$. Then $F(1) = 1 = F(\alpha) + F(\beta)$. This means that $F(\alpha) = \alpha$ and $F(\beta) = \beta$. Since $\mathrm{Ker}\,(F - I) = \mathbb{F}_p$ we must have $\alpha \in \mathbb{F}_p$. Therefore $\alpha = 0$ and $\beta = 1 \in \mathrm{Im}(\varphi)$. □

By Theorem 4.9.3 we know that a polynomial is irreducible if and only if the two conditions $\mathrm{Ker}\,(F) = 0$ and $\mathrm{Ker}\,(F - I) = \mathbb{F}_p$ are satisfied. If one of these conditions fails then we have seen how to extract a non-trivial factor in f. This procedure is called Berlekamp's algorithm ([3]). For small prime numbers p it is very efficient for finding non-trivial factors.

Remark 4.9.4 If f is divisible by the square π^2 of an irreducible polynomial $\pi \in \mathbb{F}_p[X]$ then one can find a non-trivial factor of f by computing $\gcd(f, D(f))$. This is a consequence of Lemma 4.3.8.

4.10 Exercises

1. Let R be a commutative ring and let $F = F(R, R)$ be the set of functions $f : R \to R$. Functions in F can be added and multiplied by borrowing the operations from R:

$$(f + g)(x) = f(x) + g(x)$$
$$(fg)(x) = f(x)g(x).$$

   (i) Prove that F is a commutative ring.
   (ii) Let I denote the identity function $I(r) = r$ in F. Prove that the map $\varphi : R[X] \to F$ given by

$$\varphi(a_n X^n + \cdots + a_1 X + a_0) = a_n I^n + \cdots + a_1 I + a_0$$

   is a ring homomorphism.
   (iii) Give an example showing that φ in general is not injective (hint: try $R = \mathbb{F}_2$). Find $\mathrm{Ker}\,(\varphi)$ when $R = \mathbb{F}_p$.
   The fact that φ is not injective means that one cannot in general view polynomials in $R[X]$ as R-valued functions on R.
2. Let $f, g, h \in R[X] = R[\mathbb{N}]$.
   (i) Prove that $fg = gf$.
   (ii) Prove that $f(g + h) = fg + fh$.
   (iii) Prove that $f(gh) = (fg)h$ by reducing to the case $h = cX^m$, where $c \in R$ and $m \in \mathbb{N}$.

3. Let $f, g \in R[X] \setminus \{0\}$ with $f + g \neq 0$. Prove that

$$\deg(f + g) \leq \max(\deg(f), \deg(g)).$$

4. Prove that a monic polynomial $q \in R[X]$ is not a zero divisor. Prove also that $qf = qg$ implies $f = g$, where $f, g \in R[X]$.

5. Prove that $R[X]$ is a domain if R is a domain.

6. Let R be the ring of functions $f : \mathbb{N} \to \mathbb{Z}/2\mathbb{Z}$. Recall that $(f + g)(n) = f(n) + g(n)$ and $(fg)(n) = f(n)g(n)$, where $f, g \in R$. Prove that the polynomial $X^2 - X \in R[X]$ has infinitely many roots.

7. Show that there are infinitely many irreducible polynomials in $\mathbb{F}_p[X]$, where p is a prime number (hint: look at the proof of Theorem 1.8.2).

8. Let R be a unique factorization domain and K the field of fractions $Q(R)$ of R. Suppose that $\alpha = a/s \in K$ and that a and s have no associated prime divisors. Prove that $s \mid a_n$ and $a \mid a_0$ if α is a root in the polynomial

$$a_n X^n + \cdots + a_1 X + a_0 \in K[X],$$

where $a_n, \ldots, a_1, a_0 \in R$. Use this to prove that a real number $\zeta \in \mathbb{R} \setminus \mathbb{Z}$, which is a root in a monic polynomial with integer coefficients, cannot be rational.

9. We let $D : R[X] \to R[X]$ denote the derivative introduced in subsection 4.3.1.
   (i) Prove that $D(f + g) = D(f) + D(g)$, where $f, g \in R[X]$.
   (ii) Prove that $D(\lambda f) = \lambda D(f)$, where $\lambda \in R$ and $f \in R[X]$.

10. Show that $\Phi_p(X) = X^{p-1} + \cdots + X + 1$, where p is a prime number.

11. Show that $\Phi_{p^r}(X) = \Phi_p(X^{p^{r-1}})$, where p is a prime number.

12. Prove that $\Phi_{2n}(X) = \Phi_n(-X)$, if n is odd and > 1.

13. Let $f = \Phi_8(X) = X^4 + 1$.
   (i) Prove that f is reducible in $\mathbb{F}_p[X]$ for $p \equiv 1 \pmod 4$.
   (ii) Suppose that $p \equiv 3 \pmod 8$. Prove that we may find $a \in \mathbb{F}_p$ with $a^2 = -2$. Prove for this a that $f = (X^2 + aX - 1)(X^2 - aX - 1)$ in $\mathbb{F}_p[X]$.
   (iii) Suppose that $p \equiv 7 \pmod 8$. Prove that we may find $a \in \mathbb{F}_p$ with $a^2 = 2$. Prove for this a that $f = (X^2 + aX + 1)(X^2 - aX + 1)$ in $\mathbb{F}_p[X]$.
   (iv) Conclude that f is reducible in $\mathbb{F}_p[X]$ for every prime number p.

14. Let $n \in \mathbb{N}$ with $n > 1$.
   (i) Prove that the set of nth roots of unity is a subgroup of $(\mathbb{C}^*, \cdot)$ isomorphic to $\mathbb{Z}/n\mathbb{Z}$.

(ii) Use Lemma 4.4.1 to prove that $\mathbb{Z}/n\mathbb{Z}$ contains $\varphi(n)$ elements of order n.

15. Let $\varphi : R \to S$ be a ring homomorphism. Prove that $\varphi' : R[X] \to S[X]$ given by

$$\varphi'(a_0 + a_1 X + \cdots + a_n X^n) = \varphi(a_0) + \varphi(a_1)X + \cdots + \varphi(a_n)X^n$$

is a ring homomorphism.

16. Let R be a domain. Prove that a finite subgroup of R^* is cyclic.

17. Find a generator of the cyclic group $\mathbb{F}_{17}^*$.

18. Let G be a finite subgroup of $\mathbb{C}^*$. Prove, without using Theorem 4.5.3, that G is cyclic.

19. Prove that $\mathbb{R}^*$ is not a cyclic group.

20. Prove that a natural number p is a prime number if and only if

$$a^{p-1} \equiv 1 \,(\mathrm{mod}\ p),$$
$$a^{(p-1)/q} \not\equiv 1 \,(\mathrm{mod}\ p) \text{ for every prime } q \mid p - 1$$

for some integer a.

21. Let p be a prime number $\neq 2$, $a \in \mathbb{N}$ a primitive root modulo p and $G = (\mathbb{Z}/p^2\mathbb{Z})^*$.
    (i) Prove that $\mathrm{ord}\,_G([a]) = p - 1$ or $p(p - 1)$.
    (ii) Suppose that $a^{p-1} \equiv 1 \,(\mathrm{mod}\ p^2)$. Prove that $r^{p-1} = 1 + tp$, where $r = a + p$ and $p \nmid t$.
    (iii) Prove that $\mathrm{ord}\,_G([a + p]) = p(p - 1)$ if $\mathrm{ord}\,_G([a]) = p - 1$.
    (iv) Conclude that $(\mathbb{Z}/p^2\mathbb{Z})^*$ is a cyclic group.
    (v) Suppose that $a^{p-1} = 1 + tp$, where $p \nmid t$. Prove that

    $$a^{p^{m-1}(p-1)} = 1 + t_m p^m$$

    where $m > 1$ and $p \nmid t_m$.
    (vi) Prove that $(\mathbb{Z}/p^m\mathbb{Z})^*$ is a cyclic group if $m \geq 1$.

22. Let a be a primitive root modulo the prime number $p > 2$. Show that

$$a^{(p-1)/2} \equiv -1 \,(\mathrm{mod}\ p).$$

23. Let p be a prime number.
    (i) Suppose that s is a non-zero natural number such that $p \mid 10^s - 1$. Prove that the period length of $1/p$ is $\leq s$ (hint: write $1/p = x/10^s + 1/10^s \cdot 1/p$ for a natural number $0 \leq x < 10^s$).
    (ii) Prove that the period length of $1/p$ is $\leq p - 1$.
    (iii) Prove that the period length of $1/p$ is the order of $[10]$ in $\mathbb{F}_p^*$.

24. Let p be an odd prime number and let $\alpha = [X] \in R = \mathbb{F}_p[X]/\langle X^4 + 1 \rangle$.
    (i) Prove that α is a primitive eighth root of unity in R.
    (ii) Let $y = \alpha + \alpha^{-1}$. Prove that $y^2 = 2$ and that $y^p = \alpha^p + \alpha^{-p}$.
    (iii) Prove that $y^p = y$ if $p \equiv 1, 7 \,(\mathrm{mod}\ 8)$ and that $y^p = -y$ if
        $p \equiv 3, 5 \,(\mathrm{mod}\ 8)$.
    (iv) Use the facts on $y \in R$ developed earlier in this exercise to prove that
        $(\frac{2}{p}) = 1$ if $p \equiv 1, 7 \,(\mathrm{mod}\ 8)$ and $(\frac{2}{p}) = -1$ if $p \equiv 3, 5 \,(\mathrm{mod}\ 8)$.
25. Compute a greatest common divisor d of $f = X^7 + X^6 + X^2 + X + 1$
    and $g = X^7 + X^5 + X^4 + X^2 + 1$ in $\mathbb{F}_2[X]$ along with $\lambda, \mu \in \mathbb{F}_2[X]$
    such that $\lambda f + \mu g = d$.
26. Let $R = \mathbb{F}_3[X]$.
    (i) Show that $X^2 + 1$, $X^2 + X + 2$ and $X^2 + 2X + 2$ are the only monic
        irreducible polynomials of degree 2 in R.
    (ii) Show that if a polynomial $f \in R$ of degree 4 or 5 with no roots is
        reducible then there is a monic irreducible polynomial of degree 2
        dividing f.
    (iii) Show that $X^5 - X + 1$ is an irreducible polynomial in R and that
        $L = R/\langle X^5 - X + 1 \rangle$ is a field with 243 elements. Let $\alpha = [X]$.
        Find an element $\gamma \in L$ such that $\alpha \gamma = 1$ in L.
27. Show that if a polynomial $f \in \mathbb{C}[X]$ is irreducible then $\deg(f) = 1$.
28. Let $R = \mathbb{F}_2[X]$.
    (i) Show that $X^5 + X + 1$ is not an irreducible polynomial in R.
    (ii) Show that $X^4 + X + 1$ is an irreducible polynomial in R.
    (iii) Show that $L = R/\langle X^4 + X + 1 \rangle$ is a field with 16 elements.
    (iv) Show that L^* is a cyclic group and that $L^* = \langle \alpha \rangle$, where $\alpha = [X]$.
29. Let $L = \mathbb{F}_2[X]/\langle X^3 + X + 1 \rangle$.
    (i) Show that $|L| = 8$.
    (ii) Write down the seven elements in L^*. Show by explicit computation
        that their product is -1.
    (iii) Let K be a finite field with N elements. Show that the polynomial

$$X^{N-1} - 1 \in K[X]$$

    is a product of $N - 1$ polynomials of degree 1 with non-zero
    constant coefficient.
    (iv) Let π be the product of the elements in K^*. Show that $\pi = -1$.
30. Let $R = \mathbb{F}_2[X]/\langle X^3 + 1 \rangle$ and $\alpha = [X] \in R$.
    (i) Show that $(X^2 + X + 1)(X + 1)$ is an irreducible factorization of
        $X^3 + 1$ in $\mathbb{F}_2[X]$.
    (ii) Show that $|R| = 8$ and $(\alpha^2 + \alpha + 1)(\alpha + 1) = 0$.

(iii) Show that $\alpha^2 + \alpha + 1$, $\alpha + 1$, $\alpha^2 + \alpha$, $\alpha^2 + 1$ cannot be units in R.

(iv) Show that R^* is cyclic of order 3.

31. Let $R = \mathbb{F}_2[X]$.

(i) Show that $X^2 + X + 1$ is the only irreducible polynomial of degree 2 in R.

(ii) Show that $X^3 + X + 1$ and $X^3 + X^2 + 1$ are the only irreducible polynomials of degree 3 in R.

(iii) Find two distinct irreducible polynomials f and g of degree 6 in R.

(iv) Use the notation from (iii). Prove that the rings $R/\langle f \rangle$ and $R/\langle g \rangle$ are isomorphic.

32. Let $R = \mathbb{F}_2[X]$.

(i) Show that $X - 1 \mid X^7 - 1$ and compute the polynomial $f = (X^7 - 1)/(X - 1)$. Prove that $R/\langle f \rangle$ is a ring with 64 elements.

(ii) List the irreducible polynomials in R of degree 3 and write f as a product of irreducible polynomials.

(iii) Prove that $R/\langle f \rangle$ is not a field.

33. Construct a field with eight elements.

34. Give an example of an infinite field of characteristic $p > 0$.

35. List the monic irreducible polynomials of degree 3 in $\mathbb{F}_3[X]$.

36. List the monic irreducible polynomials of degree 4 in $\mathbb{F}_2[X]$.

37. Suppose that the ring R contains the field F as a subring. Prove that R is a vector space over F using the multiplication in R (see Appendix B).

38. Let K be a finite field with p^n elements and $L \subseteq K$ a subfield with p^m elements

(i) Prove that $m \mid n$ (see Exercise 4.37).

(ii) Suppose that $r \mid s$, where $r, s \in \mathbb{N}$. Prove that

$$X^{p^r} - X \mid X^{p^s} - X$$

in $\mathbb{Z}[X]$.

(iii) Prove that K contains a subfield with p^m elements if $m \mid n$ by showing explicitly that

$$\{x \in K \mid x^{p^m} = x\}$$

is a subfield of K with p^m elements.

39. Show that there are 440 monic irreducible polynomials of degree 3 in $\mathbb{F}_{11}[X]$.

40. Show that there are 804076 monic irreducible polynomials of degree 6 in $\mathbb{F}_{13}[X]$.

41. Prove Lemma 4.8.10 and apply it to factor
    $X^5 + X^4 + X^3 + 2X + 2 \in \mathbb{F}_3[X]$.
42. Use Berlekamp's algorithm to find a prime factorization of
    $f = X^6 + X^5 + X^4 + X^3 + X^2 + X + 1 \in \mathbb{F}_2[X]$. Compare with
    Exercise 4.32.
43. Consider the nth cyclotomic polynomial Φ_n in $\mathbb{F}_p[X]$, where $p \nmid n$. Let π
    be an irreducible polynomial of degree d in $\mathbb{F}_p[X]$ that divides Φ_n. Put
    $m = \mathrm{ord}([p])$ in the group $(\mathbb{Z}/n\mathbb{Z})^*$ and let

    $$\alpha = [X] \in L = \mathbb{F}_p[X]/\langle \pi \rangle.$$

    (i) Prove that L is a field with p^d elements and that α is a primitive nth
        root of unity in L. Show that this implies that $p^d \equiv 1 \pmod{n}$.
    (ii) Prove that $L' = \{\xi \in L \mid \xi^{p^m} = \xi\}$ is a subfield of L. Prove that
        $\alpha \in L'$ and that $L' = L$. Conclude that $p^d \le p^m$.
    (iii) Prove that Φ_n is a product of distinct irreducible polynomials of
        degree m in $\mathbb{F}_p[X]$.
    (iv) Prove that Φ_n is irreducible in $\mathbb{F}_p[X]$ if and only if $[p]$ generates
        $(\mathbb{Z}/n\mathbb{Z})^*$.
44. Let $f, g \in \mathbb{Q}[X] \setminus \{0\}$. Prove that if f is an irreducible polynomial and
    $f(z) = g(z) = 0$ for some complex number $z \in \mathbb{C}$ then $f \mid g$ in $\mathbb{Q}[X]$.
45. **(HOF)** This exercise is a guided tour of the proof that cyclotomic
    polynomials are irreducible as polynomials in $\mathbb{Q}[X]$. Needless to say, this
    result goes back to Gauss. Let $n \ge 1$ and f an irreducible monic
    polynomial dividing Φ_n in $\mathbb{Q}[X]$.
    (i) Consider f as a polynomial in $\mathbb{C}[X]$. Prove that $f(\zeta) = 0$ for some
        primitive nth root of unity ζ.
    (ii) Prove that the fact that every primitive nth root of unity is a root in f
        if $f(\zeta) = 0$ implies that $f(\zeta^p) = 0$, where ζ is a primitive nth root
        of unity and p is a prime number not dividing n.
    (iii) Let f and g be monic polynomials in $\mathbb{Q}[X]$. Prove that $f, g \in \mathbb{Z}[X]$
        if $fg \in \mathbb{Z}[X]$.
    (iv) Prove that $f \mid X^n - 1$ in $\mathbb{Q}[X]$ and write

    $$X^n - 1 = f(X)g(X).$$

    Prove that $f, g \in \mathbb{Z}[X]$.
    (v) Let ζ be a primitive nth root of unity such that $f(\zeta) = 0$. Suppose
        that p is a prime number not dividing n and $f(\zeta^p) \ne 0$. Prove that ζ
        is a root in $g(X^p)$ and that $f(X) \mid g(X^p)$ (see Exercise 4.44).
        Write $g(X^p) = f(X)h(X)$ for $h(X) \in \mathbb{Q}[X]$. Consider the

corresponding polynomials $\bar{f}, \bar{g}, \bar{h} \in \mathbb{F}_p[X]$. Prove that $\bar{g}(X^p) = \bar{g}(X)^p$ and that an irreducible polynomial $\pi \in \mathbb{F}_p[X]$ dividing $\bar{f}$ must divide $\bar{g}$.

(vi) Why is it impossible for $\bar{f}, \bar{g} \in \mathbb{F}_p[X]$ to have a common prime divisor when $p \nmid n$?

(vii) Prove that $f(\zeta) = 0$ implies that $f(\zeta^p) = 0$, where ζ is a primitive nth root of unity and $p \nmid n$. Show that this implies that f equals Φ_n and therefore that Φ_n is irreducible.

5 Gröbner bases

A symmetric function $f : \mathbb{R}^2 \to \mathbb{R}$ is a function satisfying $f(X, Y) = f(Y, X)$ for every $(X, Y) \in \mathbb{R}^2$. Simple examples of symmetric functions are $s_1(X, Y) = X + Y$ and $s_2(X, Y) = XY$. Polynomials in X and Y are functions built from addition and multiplication of the variables X and Y, such as $f(X, Y) = X^5Y + X + Y$. The polynomial $f(X, Y) = X^2 + Y^2$ is an example of a polynomial that is a symmetric function. We call it a symmetric polynomial. A special case of a classical result due to Newton (1643–1727) says that every symmetric polynomial is a polynomial in s_1 and s_2. For example,

$$X^2 + Y^2 = (X + Y)^2 - 2XY = s_1^2 - 2s_2$$

and

$$X^3 + Y^3 = (X + Y)^3 - 3(X + Y)XY = s_1^3 - 3s_1s_2.$$

You may want to continue the list with $X^4 + Y^4$ or to wait until you have digested the rudiments of the theory of Gröbner bases and can understand "Newton revisited" (Section 5.5). In this chapter we will develop the theory of Gröbner bases in polynomial rings in several variables. The original impetus for this recent development of algebra was the desire to solve equations. Systems of linear equations such as

$$5x + y + z = 17,$$
$$x + y - z = 1,$$
$$x + y + z = 9$$

can be solved using Gaussian elimination. However, many problems lead to systems of non-linear equations, such as

$$y^2 - x^3 + x = 0,$$
$$y^3 - x^2 = 0,$$

186

where the variables occur with powers greater than 1. The theory of Gröbner bases is a far-reaching generalization of Gaussian elimination. It can be applied for solving systems of non-linear (polynomial) equations such as above. Gröbner bases were invented independently by Buchberger (1942–) and Hironaka (1931–) in the sixties. Hironaka used the term "standard bases" in connection with his work on resolution of singularities in algebraic geometry (1964). Buchberger used the term Gröbner bases in his Ph.D. thesis (1966), in honor of his advisor W. Gröbner (1899–1980) . In accordance with most modern mathematical literature we will use this term. Gröbner bases have some remarkable (mathematical) properties and turn out to be useful also in areas not confined to the world of mathematics, for example in optimization, robotics and theoretical computer science.

5.1 Polynomials in several variables

So far we have only encountered and defined polynomials in one variable. We need to define formally polynomials in more than one variable. Fortunately it is very easy to modify our formal construction of polynomials in one variable. Recall that the ring of polynomials $R[X]$ with coefficients in a (commutative) ring R was defined as

$$R[X] = R[\mathbb{N}] = \{f : \mathbb{N} \to R \mid f(n) = 0, \ n \gg 0\}$$

with obvious addition and not so obvious multiplication (see Section 4.1). A polynomial $f \in R[X]$ in one variable can be expressed in the usual notation as

$$f = a_n X^n + \cdots + a_1 X + a_0, \qquad a_i \in R,$$

and addition and multiplication coincide with well known operations (but with coefficients in an arbitrary ring). Polynomials in several variables should correspond to algebraic expressions like $X^2 + XY + Y + Y^3 + X^5$ (in the case of two variables X and Y). We define the polynomial ring $R[X_1, \ldots, X_n]$ in n variables $X_1, \ldots, X_n$ as

$$R[X_1, \ldots, X_n] = R[\mathbb{N}^n] = \{f : \mathbb{N}^n \to R \mid f(v) = 0, \ |v| \gg 0\},$$

where $v = (v_1, \ldots, v_n) \in \mathbb{N}^n$ and $|v| = v_1 + \cdots + v_n$. A polynomial $f \in R[X_1, \ldots, X_n]$ is the same as a function $f : \mathbb{N}^n \to R$ that is non-zero for only finitely many $v \in \mathbb{N}^n$. We let $X^v \in R[\mathbb{N}^n]$ denote the function given by

$$X^v(w) = \begin{cases} 1 & \text{if } v = w, \\ 0 & \text{if } v \neq w. \end{cases}$$

With this notation, every polynomial $f \in R[\mathbb{N}^n]$ can be written as a (finite) sum

$$f = \sum_{v \in \mathbb{N}^n} a_v X^v,$$

where $a_v \in R$ (an element $r \in R$ is identified with the function mapping the zero vector to r and everything else to $0 \in R$). If $f, g \in R[\mathbb{N}^n]$ we define $f + g$ by $(f + g)(v) = f(v) + g(v)$ and fg by the (finite) sum

$$(fg)(v) = \sum_{v_1 + v_2 = v} f(v_1) g(v_2),$$

where $v_1, v_2 \in \mathbb{N}^n$. The complete proof that $R[\mathbb{N}^n]$ is a ring with these compositions is left to the reader (see Exercise 5.1), as it is very similar to the one-variable case. We note that $0 \in R$ is the neutral element for $+$ and that the function $X^{(0,0,\dots,0)}$, mapping the zero vector in $\mathbb{N}^n$ to $1 \in R$ and everything else to 0, is the neutral element for multiplication. In the notation $R[X_1, \dots, X_n]$ for $R[\mathbb{N}^n]$, X_1 refers to $X^{(1,0,\dots,0)}$, X_2 to $X^{(0,1,0,\dots,0)}$, $\dots$ and X_n to $X^{(0,\dots,0,1)}$.

A *term* is a polynomial $r X^v \in R[\mathbb{N}^n]$, where $r \in R \setminus \{0\}$ is called the *coefficient*.

Example 5.1.1   The formal definition of polynomials in several variables is a precise mathematical model for polynomial expressions in variables $X, Y, Z, \dots$. Be sure that you understand how to go from the formal expressions to the "real-world" expressions in $X, Y, Z, \dots$ and back. As an example, let

$$f = 2X^{(0,0,0)} + 2X^{(1,0,3)} + X^{(2,1,0)} - X^{(0,1,1)} + 3X^{(1,1,1)} \in \mathbb{Z}[\mathbb{N}^3].$$

Translating X to $X^{(1,0,0)}$, Y to $X^{(0,1,0)}$ and Z to $X^{(0,0,1)}$ we get

$$f = 2 + 2XZ^3 + X^2Y - YZ + 3XYZ \in \mathbb{Z}[X, Y, Z]$$

as the corresponding polynomial expression in X, Y and Z. Multiplying polynomials in several variables corresponds to the natural way of multiplying and collecting terms, e.g.

$$(X + 2Y - Z)(X + Y - Z) = X^2 + XY - XZ + 2XY + 2Y^2 - 2YZ$$
$$- XZ - YZ + Z^2 = X^2 + 3XY - 2XZ + 2Y^2 - 3YZ + Z^2.$$

5.1.1 Term orderings

In one variable it is natural that a term like X^5 is bigger than X^3. In more than one variable there is no obvious way of ordering the individual terms. In two variables, how should we compare terms like X^2Y and X^3? This is formalized in the notion of a term ordering. The price we pay for comparing terms in more than one variable is that there are infinitely many natural ways of doing it (see Remark 5.1.4). Before reading on, you should consult Appendix A for the definitions of a partial and a total ordering on a set.

Definition 5.1.2   The set $\mathbb{N}^n$ of n-tuples of natural numbers carries a natural component-wise addition $+$ with zero vector $0 = (0, \dots, 0)$. A partial ordering $\leq$ on $\mathbb{N}^n$ is called a *term ordering* if

(i) $\leq$ is a total ordering,

(ii) $0 \leq v$,

(iii) $v_1 \leq v_2 \Rightarrow v_1 + v \leq v_2 + v$

for every $v, v_1, v_2 \in \mathbb{N}^n$.

Example 5.1.3   We will give a few examples of term orderings.

(1) A term ordering on $\mathbb{N} = \mathbb{N}^1$ has to be the usual total ordering on $\mathbb{N}$ (why?).

(2) Define the *lexicographic ordering* $\leq_{\mathrm{lex}}$ on $\mathbb{N}^n$ by

$$(v_1, \dots, v_n) \leq_{\mathrm{lex}} (w_1, \dots, w_n)$$

if one of the following applies:

$(v_1 < w_1)$ or

$(v_1 = w_1)$ and $(v_2 < w_2)$ or

$(v_1 = w_1)$ and $(v_2 = w_2)$ and $(v_3 < w_3)$ or $\dots$

$(v_1 = w_1)$ and $(v_2 = w_2)$ and $\dots$ and $(v_n = w_n)$.

This is nothing but "alphabetic" ordering on tuples of natural numbers; for example, $(1, 2, 3) \geq_{\mathrm{lex}} (1, 1, 3)$ since $2 > 1$ and $(4, 5, 1) \leq_{\mathrm{lex}} (4, 5, 3)$ since $1 < 3$.

(3) Let $|v| = v_1 + v_2 + \dots + v_n$, where $v = (v_1, \dots, v_n) \in \mathbb{N}^n$. Define the *graded lexicographic ordering* by $v \leq_{\mathrm{grlex}} w$ if $|v| < |w|$ or $|v| = |w|$ and $v \leq_{\mathrm{lex}} w$. Notice that, for example, $(1, 2, 3) \geq_{\mathrm{grlex}} (2, 1, 1)$ (since $1 + 2 + 3 > 2 + 1 + 1$) but $(1, 2, 3) \leq_{\mathrm{lex}} (2, 1, 1)$.

You should check immediately that $\leq_{\text{lex}}$ and $\leq_{\text{grlex}}$ are partial orderings and that they satisfy the three rules defining a term ordering (see Exercise 5.7).

A fruitful way of studying term orderings is through a little geometry. For a vector $v \in \mathbb{R}^n$ of real numbers ≥ 0 one can construct a term ordering $\leq_v$ on $\mathbb{N}^n$ defined as $u_1 \leq_v u_2$ if and only if

$$v \cdot u_1 < v \cdot u_2 \quad \text{or} \quad (v \cdot u_1 = v \cdot u_2 \text{ and } u_1 \leq_{\text{lex}} u_2), \quad (5.1)$$

where $u_1, u_2 \in \mathbb{N}^n$ and $\cdot$ refers to the usual inner product on $\mathbb{R}^n$ (see Exercise 5.8).

Remark 5.1.4   There is a fundamental difference between $\mathbb{N}$ and $\mathbb{N}^2$. On $\mathbb{N}$ there is only one term ordering. On $\mathbb{N}^2$ there are infinitely many (in fact uncountably many). Let $\leq_r$ denote the term ordering on $\mathbb{N}^2$ given by the vector $(1, r)$ as in (5.1), where r is a positive real number. If $s \neq r$ is another positive real number, we may find $v \in \mathbb{Z}^2$ such that $(1, r) \cdot v > 0$ and $(1, s) \cdot v < 0$. You can see this by drawing the lines through $(0, 0)$ orthogonal to $(1, r)$ and $(1, s)$. Any point with integer coordinates between the two diagonal lines will do.

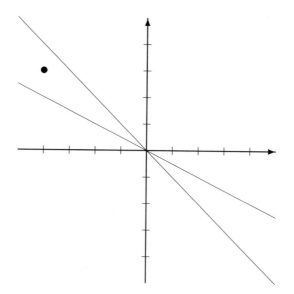

A vector in $\mathbb{Z}^2$ can always be written as the difference of two vectors in $\mathbb{N}^2$ (e.g. $(1, -1) = (1, 0) - (0, 1), (-1, -1) = (0, 0) - (1, 1)$ and $(1, 1) = (1, 1) - (0, 0)$). Write $v = v_1 - v_2$, where $v_1, v_2 \in \mathbb{N}^2$. Then $v_1 \geq_r v_2$ but $v_1 \leq_s v_2$. Thus for every positive real number r we have defined a term ordering $\leq_r$ such that if s is another positive real number then $\le_r \ne \le_s$. This shows that there are infinitely (uncountably) many term orderings on $\mathbb{N}^2$.

For a given vector $v \in \mathbb{N}^n$ we let

$$v + \mathbb{N}^n = \{v + w \mid w \in \mathbb{N}^n\}.$$

We will need the following crucial result, known as Dickson's lemma (L. E. Dickson (1874–1954)). It originally appeared in a paper on number theory ([7], Lemma A).

Lemma 5.1.5 (Dickson) *Let S be a subset of $\mathbb{N}^n$. Then there is a finite set of vectors $v_1, \ldots, v_r \in S$ such that*

$$S \subseteq (v_1 + \mathbb{N}^n) \cup \cdots \cup (v_r + \mathbb{N}^n).$$

Example 5.1.6 The idea of the proof is really quite simple and comes from the case of the subsets of $\mathbb{N}^2$. In the figure below we show a certain infinite subset $S \subseteq \mathbb{N}^2$ (extended infinitely in the positive x- and y- directions).

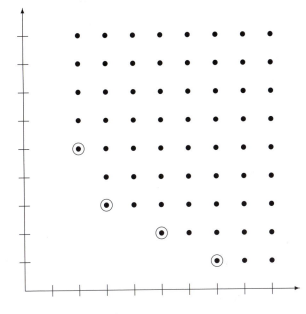

The marked points are the interesting points for the subset S, in that

$$S \subseteq ((2, 5) + \mathbb{N}^2) \cup ((3, 3) + \mathbb{N}^2) \cup ((5, 2) + \mathbb{N}^2) \cup ((7, 1) + \mathbb{N}^2).$$

Proof of Lemma 5.1.5. The proof proceeds by induction on n. If $n = 1$ and $S \subseteq \mathbb{N}$ is a subset, we let s be the first element in S. Then $S \subseteq s + \mathbb{N}$. Suppose now for the induction step that $n > 1$ and we know that Lemma 5.1.5 is true for $m < n$. Let $\pi : \mathbb{N}^n \to \mathbb{N}^{n-1}$ denote the map given by

$$\pi(x_1, x_2, \ldots, x_n) = (x_2, \ldots, x_n).$$

Using the induction hypothesis on the subset

$$\pi(S) = \{\pi(s) \mid s \in S\} \subseteq \mathbb{N}^{n-1}$$

we get the existence of $s_1, \ldots, s_r \in S$ such that

$$\pi(S) \subseteq (\pi(s_1) + \mathbb{N}^{n-1}) \cup \cdots \cup (\pi(s_r) + \mathbb{N}^{n-1}).$$

It is in general not true that $S \subseteq (s_1 + \mathbb{N}^n) \cup \cdots \cup (s_r + \mathbb{N}^n)$ (you can see this in Example 5.1.6). We need more vectors in S.

Let M be the largest number occurring as a first coordinate in our vectors $s_1, \ldots, s_r$. Define

$$S_i = \{s \in S \mid \text{the first coordinate of } s = i\} \qquad \text{for } 0 \le i < M$$

and

$$S_{\ge M} = \{s \in S \mid \text{ the first coordinate of } s \text{ is } \ge M\}.$$

Then $S = S_0 \cup \cdots \cup S_{M-1} \cup S_{\ge M}$ and

$$S_{\ge M} \subseteq (s_1 + \mathbb{N}^n) \cup \cdots \cup (s_r + \mathbb{N}^n).$$

Since the first coordinate of the vectors in S_i is fixed, we can identify S_i with a subset of $\mathbb{N}^{n-1}$ and by induction find finitely many vectors $s_1^i, \ldots, s_{r_i}^i \in S_i$ such that

$$S_i \subseteq \left(s_1^i + \mathbb{N}^n\right) \cup \cdots \cup \left(s_{r_i}^i + \mathbb{N}^n\right).$$

Gathering up these finitely many vectors for $S_0, \ldots, S_{M-1}$ and throwing in the vectors $s_1, \ldots, s_r$ we get the result. $\square$

Make sure you understand how the proof of Lemma 5.1.5 works for the subset $S \subseteq \mathbb{N}^2$ in Example 5.1.6.

Corollary 5.1.7 *A term ordering $\le$ on $\mathbb{N}^n$ is a well ordering.*

Proof. Let $S \subseteq \mathbb{N}^n$ be a non-empty subset. By Lemma 5.1.5 there are finitely many elements $v_1, \ldots, v_r \in S$ such that

$$S \subseteq (v_1 + \mathbb{N}^n) \cup \cdots \cup (v_r + \mathbb{N}^n).$$

If $v \in v_i + \mathbb{N}^n$ then $v = v_i + w$ for some $w \in \mathbb{N}^n$. This implies that $v - v_i \in \mathbb{N}^n$. Since $v - v_i \geq 0$ by Definition 5.1.2(ii), it follows that $v = (v - v_i) + v_i \geq v_i$ by Definition 5.1.2(iii). This means that the smallest element among $v_1, \ldots, v_r$ will be the smallest element in S, showing that $\leq$ is a well ordering. $\qquad\square$

5.2 The initial term of a polynomial

Definition 5.2.1   Let

$$f = \sum_{v \in \mathbb{N}^n} a_v X^v$$

be a non-zero polynomial in $R[\mathbb{N}^n]$ and $\leq$ a term order on $\mathbb{N}^n$. The initial term of f with respect to $\leq$ is defined as

$$\mathrm{in}_\leq(f) = a_w X^w,$$

where $w = \max_\leq \{v \in \mathbb{N}^n \mid a_v \neq 0\}$ (see Definition A.3.6 for the definition of $\max_\leq$). In an abuse of notation we will sometimes compare two terms and write $a X^u \leq b X^v$ if $u \leq v$.

Example 5.2.2   Let $f = X^2 + XY + Y + Y^3 + X^5 \in \mathbb{Z}[X, Y]$, where X corresponds to $X^{(1,0)}$ and Y to $X^{(0,1)}$ in $\mathbb{Z}[\mathbb{N}^2]$. This means that

$$f = X^{(2,0)} + X^{(1,1)} + X^{(0,1)} + X^{(0,3)} + X^{(5,0)} \in \mathbb{Z}[\mathbb{N}^2].$$

Putting $\leq \ = \ \leq_{\mathrm{lex}}$ (Example 5.1.3), we obtain

$$(5, 0) \geq (2, 0) \geq (1, 1) \geq (0, 3) \geq (0, 1).$$

In the ordering $\leq$ one should write $f = X^5 + X^2 + XY + Y^3 + Y$. The initial term of f is therefore $\mathrm{in}_\leq(f) = X^5$.

Remark 5.2.3   Let R be a domain and f, g non-zero polynomials in $R[X_1, \ldots, X_n]$. Then $\mathrm{in}_\leq(fg) = \mathrm{in}_\leq(f)\,\mathrm{in}_\leq(g)$ (see Exercise 5.11). This formula is the analogue of $\deg(fg) = \deg(f) + \deg(g)$ in one variable (see Proposition 4.2.2),

5.3 The division algorithm

In several variables there is an analogue of division with remainder (Proposition 4.2.4). Now everything is with respect to a fixed term ordering (in the case of one variable, there is only one term ordering; in more than one variable there are infinitely many (Remark 5.1.4)). The proof of the following proposition is based on the division algorithm in several variables. This algorithm is very similar to the one-variable algorithm given in the proof of Proposition 4.2.4. In order not to separate the algorithm from its mathematical surroundings it is embedded in the proof. To learn the algorithm and prove its correctness you will have to read through the proof and immerse yourself in several examples and exercises. We will assume for the rest of this chapter that R is a domain.

Proposition 5.3.1 (The division algorithm) *Fix a term ordering $\leq$ on $\mathbb{N}^n$. Let $f \in R[X_1, \ldots, X_n] \setminus \{0\}$ and suppose that $f_1, \ldots, f_m \in R[X_1, \ldots, X_n]$ is a sequence of non-zero polynomials. Then there exist $a_1, \ldots, a_m, r \in R[X_1, \ldots, X_n]$ such that*

$$f = a_1 f_1 + \cdots + a_m f_m + r$$

and either $r = 0$ or none of the terms in r is divisible by $\mathrm{in}_\leq(f_1), \ldots, \mathrm{in}_\leq(f_n)$. Furthermore, $\mathrm{in}_\leq(a_i f_i) \leq \mathrm{in}_\leq(f)$ if $a_i f_i \neq 0$.

Proof.   The proof is basically a correctness proof of the division algorithm for polynomials in several variables. This algorithm is similar to the algorithm in one variable as described in the proof of Proposition 4.2.4. You should compare the two. Here is the division algorithm in several variables. To begin put $a_1 := 0, \ldots, a_m := 0, r := 0$ and $s := f$ giving

$$f = a_1 f_1 + \cdots + a_m f_m + (r + s). \tag{5.2}$$

This expression will serve as an invariant throughout the algorithm. Proceed as follows in successive steps of the algorithm. If $s = 0$ we are done. If not, there are two cases. If $\mathrm{in}_\leq(s)$ is divisible by some $\mathrm{in}_\leq(f_i)$ then pick the *smallest i* with this property and let

$$s := s - \frac{\mathrm{in}_\leq(s)}{\mathrm{in}_\leq(f_i)} f_i,$$
$$a_i := a_i + \frac{\mathrm{in}_\leq(s)}{\mathrm{in}_\leq(f_i)}. \tag{5.3}$$

Notice that (5.2) still holds after the assignments in (5.3) – we have simply subtracted and added the same thing. However, if $\mathrm{in}_\leq(s)$ is not divisible by any

$\text{in}_\le(f_i)$ we add the initial term to r and subtract it from s:

$$r := r + \text{in}_\le(s),$$
$$s := s - \text{in}_\le(s). \tag{5.4}$$

Of course, after the assignments in (5.4) $r + s$ is unchanged and (5.2) still holds. If $s = 0$ we are done. If not, the initial term of s is strictly decreased after the assignment in (5.3), because $\text{in}_\le(s)\, t \lneq \text{in}_\le(s)\,\text{in}_\le(f_i)$ for a term t in f_i different from $\text{in}_\le(f_i)$. The initial term of s is also strictly decreased after the assignment in (5.4). In this way the sequence formed by $\text{in}_\le(s)$ in successive steps of the algorithm is strictly decreasing with respect to the term ordering $\le$. Since $\le$ is a well ordering by Corollary 5.1.7, such a sequence must be finite (see Lemma A.3.8). Therefore the division algorithm eventually terminates with $s = 0$. Then (5.2) is the desired expression. We have seen that $\text{in}_\le(s) \le \text{in}_\le(f)$ holds if $s \ne 0$, since s initially takes the value of f. Since $\text{in}_\le((a_i + \text{in}_\le(s)/\text{in}_\le(f_i))f_i) = \text{in}_\le(a_i \,\text{in}_\le(f_i) + \text{in}_\le(s)) \le \max(\text{in}_\le(a_i\,f_i),\ \text{in}_\le(s))$ (see Exercise 5.12) for $a_i \ne 0$ we must have $\text{in}_\le(a_i\,f_i) \le \text{in}_\le(f)$ after the assignment in (5.3). This proves that $\text{in}_\le(a_i\,f_i) \le \text{in}_\le(f)$ if $a_i\,f_i \ne 0$ in (5.2) when reaching $s = 0$. $\qquad\square$

Definition 5.3.2 Suppose that $f \in R[X_1, \ldots, X_n]$ and let $F = (f_1, \ldots, f_m)$ be a sequence of non-zero polynomials in $R[X_1, \ldots, X_n]$. We let f^F denote the remainder r coming from dividing f by F using the division algorithm.

Example 5.3.3 Let $\le\, =\, \le_{\text{lex}}$ with $Y \le X$, $f = X^4 + Y^4$, $f_1 = X^2 + Y$ and $f_2 = X^2 Y + 1$. The division algorithm is shown in the diagram below; we are trying to mimic the diagram for division of polynomials in one variable. Here, though, the result is represented by not just one polynomial but a set (a_1, a_2) of polynomials. The initial terms of f_1, f_2 and s are underlined. If the initial term of s is not divisible by either $\text{in}_\le(f_1)$ or $\text{in}_\le(f_2)$ then we transfer the initial term to the remainder r. This is indicated, for example, as $Y^4 + Y^2 \to Y^4$.

$$\begin{array}{l}
\underline{X^4} + Y^4 : (\underline{X^2} + Y,\ \underline{X^2 Y} + 1) = (X^2 - Y, 0) \\
\underline{X^4 + X^2 Y} \\
\quad -X^2 Y + Y^4 \\
\quad \underline{-X^2 Y - Y^2} \\
\qquad \underline{Y^4} + Y^2 \longrightarrow Y^4 \\
\qquad\quad \underline{Y^2} \longrightarrow Y^4 + Y^2 \\
\qquad\qquad 0
\end{array}$$

The division algorithm above shows that

$$X^4 + Y^4 = (X^2 - Y)(X^2 + Y) + Y^4 + Y^2$$

and $(X^4 + Y^4)^{(X^2+Y, X^2Y+1)} = Y^4 + Y^2$. However, suppose that we switch f_1 and f_2 (so that we divide by (f_2, f_1) instead of (f_1, f_2)). Then

$$
\begin{aligned}
\underline{X^4} + Y^4 : (\underline{X^2Y} + 1, \ \underline{X^2} + Y) &= (-1, X^2) \\
\underline{X^4 + X^2Y} & \\
- X^2Y + Y^4 & \\
\underline{-X^2Y - 1} & \\
\underline{Y^4} + 1 &\longrightarrow Y^4 \\
\underline{1} &\longrightarrow Y^4 + 1 \\
0 &
\end{aligned}
$$

This shows that

$$X^4 + Y^4 = X^2(X^2 + Y) - (X^2Y + 1) + Y^4 + 1$$

and $(X^4 + Y^4)^{(X^2Y+1, X^2+Y)} = Y^4 + 1$.

5.4 Gröbner bases

In Example 5.3.3 we saw that the remainder coming from the division algorithm depends on the order of $f_1, \ldots, f_m$ in Proposition 5.3.1. We would like to have a generating set of an ideal with the property that the remainder coming from the division algorithm is independent of the order of its elements. This is possible. Such a set of generators is called a Gröbner basis. In the rest of this chapter we will assume that R is a field denoted by k, in order to simplify the definition of a Gröbner basis (the definition for arbitrary domains is a little more complicated).

Definition 5.4.1   A set of non-zero polynomials

$$F = (f_1, \ldots, f_m) \subseteq k[X_1, \ldots, X_n]$$

is called a *Gröbner basis for an ideal* I in $k[X_1, \ldots, X_n]$ with respect to a term ordering $\leq$ if $F \subseteq I$ and, for every $f \in I \setminus \{0\}$,

$$\mathrm{in}_{\leq}(f_i) \mid \mathrm{in}_{\leq}(f)$$

for some $i = 1, \ldots, m$. The set F is called a *Gröbner basis* with respect to a term ordering $\leq$ if it is a Gröbner basis for the ideal $\langle f_1, \ldots, f_m \rangle$ with respect to $\leq$.

This definition may seem strange at first. But it is exactly to the point. As a motivating example consider the ideal $I = \langle X^2 + Y, X^2Y + 1 \rangle$ in the polynomial ring $\mathbb{Q}[X, Y]$. Recall that I consists of all the polynomials you get as "linear" combinations (see subsection 3.1.1) of $X^2 + Y$ and $X^2Y + 1$:

$$I = \{a(X, Y)(X^2 + Y) + b(X, Y)(X^2Y + 1) \mid a(X, Y), b(X, Y) \in \mathbb{Q}[X, Y]\}.$$

Thus $f = X^3 - Y + XY - X^2Y^2 \in I$ since $f = X(X^2 + Y) - Y(X^2Y + 1)$. In general, how do we decide whether a given polynomial lies in the ideal I? Here Gröbner bases and the division algorithm are very helpful.

Proposition 5.4.2 *Let* $G = (f_1, \ldots, f_m)$ *be a Gröbner basis with respect to a term ordering* $\leq$. *For a polynomial* $f \in k[X_1, \ldots, X_n]$ *we have*

$$f \in I \iff f^G = 0,$$

where $I = \langle f_1, \ldots, f_m \rangle$.

Proof. If $f^G = 0$ then $f = a_1 f_1 + \cdots + a_m f_m$ and $f \in I = \langle f_1, \ldots, f_m \rangle$. Let $f = a_1 f_1 + \cdots + a_m f_m + f^G$ be the output from the division algorithm. Taking $r = f^G$ this gives an expression for f as in Proposition 5.3.1. Clearly

$$r = f - a_1 f_1 - \cdots - a_m f_m \in I.$$

If $r \neq 0$ then there is some $\text{in}_\leq(f_i)$ dividing $\text{in}_\leq(r)$, since $(f_1, \ldots, f_m)$ was assumed to be a Gröbner basis for I. This contradicts that r is the remainder coming from division by G. Thus $r = 0$. □

Example 5.4.3 Let $F = (X^2 + Y, X^2Y + 1)$ and fix the lexicographic ordering $\leq$ on terms in $k[X, Y]$ given by $X \geq Y$. Then

$$Y^2 - 1 = Y(X^2 + Y) - (X^2Y + 1)$$

so that $Y^2 - 1 \in \langle X^2 + Y, X^2Y + 1 \rangle$. But the remainder from the division algorithm is $(Y^2 - 1)^F = Y^2 - 1$. Using Proposition 5.4.2 we see that F is not a Gröbner basis for $\langle X^2 + Y, X^2Y + 1 \rangle$. Of course this could also be checked by using the definition of a Gröbner basis. It is not too difficult to show that F is not a Gröbner basis for any term ordering (see Exercise 5.14).

Example 5.4.4   A generator (f) for a principal ideal $I \subseteq R = k[X_1, \ldots, X_n]$ is always a Gröbner basis for I. Consider a polynomial $g \in I$. Since f generates I we may find $a \in R$ such that $g = af$. Therefore $\text{in}_\leq(g) = \text{in}_\leq(a)\,\text{in}_\leq(f)$ by Remark 5.2.3 and $\text{in}_\leq(f)$ divides $\text{in}_\leq(g)$.

Corollary 5.4.5   *Let $G = (f_1, \ldots, f_m) \subseteq R = k[X_1, \ldots, X_n]$ be a Gröbner basis for the ideal $I \subseteq R$ with respect to some term ordering. Then $I = \langle f_1, \ldots, f_m \rangle$.*

Proof.   Since $f_1, \ldots, f_m \in I$ we obtain $\langle f_1, \ldots, f_m \rangle \subseteq I$. However, if $f \in I$ then $f^G = 0$ by Proposition 5.4.2 and $f = a_1 f_1 + \cdots + a_m f_m$ for suitable $a_1, \ldots, a_m \in k[X_1, \ldots, X_n]$ by the division algorithm. This proves that $I \subseteq \langle f_1, \ldots, f_m \rangle$. $\square$

Proposition 5.4.6   *Let $G = (f_1, \ldots, f_m)$ be a Gröbner basis in $R = k[X_1, \ldots, X_n]$ with respect to a term ordering $\leq$. Then the remainder r in $f = a_1 f_1 + \cdots + a_m f_m + r$ as in Proposition 5.3.1 is unique for every $f \in R$. The remainder from the division algorithm is independent of the order of the elements $f_1, \ldots, f_m$ in G.*

Proof.   Let $f \in R$ and assume we have two expressions $f = a_1 f_1 + \cdots + a_m f_m + r_1 = a_1' f_1 + \cdots + a_m' f_m + r_2$, as in Proposition 5.3.1. Then

$$r_2 - r_1 = (a_1 - a_1')f_1 + \cdots + (a_m - a_m')f_m.$$

Therefore $r_2 - r_1 \in \langle f_1, \ldots, f_m \rangle$. If $r_2 - r_1 \neq 0$ then there exists i such that $\text{in}_\leq(f_i)$ divides $\text{in}_\leq(r_2 - r_1)$. This implies that $\text{in}_\leq(f_i)$ divides a term in r_2 or r_1, which is a contradiction.

A permutation G' of the elements in G leads to an expression $f = b_1 f_1 + \cdots + b_m f_m + f^{G'}$, as in Proposition 5.3.1. This implies that $f^{G'} = f^G$, since we have just proved that the remainder in Proposition 5.3.1 is unique. $\square$

5.4.1 Hilbert's basis theorem

We will prove the existence of Gröbner bases for every ideal in $k[X_1, \ldots, X_n]$. In the late nineteenth century the German mathematician David Hilbert (1862–1943) surprised the mathematical community by showing that every ideal in a polynomial ring $k[X_1, \ldots, X_n]$ is finitely generated [13]. This is now referred to as Hilbert's basis theorem. His proof did not give explicit generators and his contemporaries were skeptical. Here is the fascinating history from the MacTutor History of Mathematics Archive.[1]

> Hilbert's first work was on invariant theory and, in 1888, he proved his famous Basis Theorem. Twenty years earlier Gordan had proved the finite basis theorem for binary forms using a highly computational approach. Attempts to generalise Gordan's work to systems with more than two variables failed since the computational difficulties were too great. Hilbert himself tried at first to follow Gordan's approach but soon realised that a new line of attack was necessary. He discovered a completely new approach which proved the finite basis theorem for any number of variables but in an entirely abstract way. Although he proved that a finite basis existed his methods did not construct such a basis. Hilbert submitted a paper proving the finite basis theorem to *Mathematische Annalen*. However, Gordan was the expert on invariant theory for *Mathematische Annalen* and he found Hilbert's revolutionary approach difficult to appreciate. He refereed the paper and sent his comments to Klein:
>
>> The problem lies not with the form ... but rather much deeper. Hilbert has scorned to present his thoughts following formal rules, he thinks it suffices that no one contradict his proof ... he is content to think that the importance and correctness of his propositions suffice. ... for a comprehensive work for the *Annalen* this is insufficient.
>
> However, Hilbert had learnt through his friend Hurwitz about Gordan's letter to Klein and Hilbert wrote himself to Klein in forceful terms:
>
>> ... I am not prepared to alter or delete anything, and regarding this paper, I say with all modesty, that this is my last word so long as no definite and irrefutable objection against my reasoning is raised.

Using the machinery of Gröbner bases, Hilbert's result follows in a remarkable way. In fact reading through the proof one tends to forget the controversies of the late nineteenth century.

Theorem 5.4.7 *Let k be a field, $\leq$ a term ordering and $I \subseteq k[X_1, \dots, X_n]$ an ideal. Then I has a Gröbner basis with respect to $\leq$.*

Proof. Let $S = \{v \in \mathbb{N}^n \mid X^v = \text{in}_{\leq}(f) \text{ for some } f \in I\} \subseteq \mathbb{N}^n$. Dickson's lemma (Lemma 5.1.5) applied to the subset S of $\mathbb{N}^n$ shows that there are finitely many $f_1, \dots, f_m \in I$ such that

$$S \subseteq (v_1 + \mathbb{N}^n) \cup \cdots \cup (v_m + \mathbb{N}^n),$$

where $X^{v_i} = \text{in}_{\leq}(f_i)$ for $i = 1, \dots, m$. Suppose that $aX^w = \text{in}_{\leq}(f)$, where $f \in I$. Then $w = v_j + v$ for a suitable $j = 1, \dots, m$ and $v \in \mathbb{N}^n$. This proves that $X^w = X^{v_j} X^v$ and therefore that $\text{in}_{\leq}(f_j) \mid \text{in}_{\leq}(f)$. This is exactly the statement that $(f_1, \dots, f_m)$ is a Gröbner basis for I. $\qquad\square$

---

[1] http://www-groups.dcs.st-and.ac.uk/~history

Corollary 5.4.8 (Hilbert)   *Let I be an arbitrary ideal in $k[X_1, \ldots, X_n]$. Then there are finitely many polynomials $f_1, \ldots, f_m \in I$ such that every polynomial $f \in I$ can be written*

$$f = a_1 f_1 + \cdots + a_m f_m$$

for suitable $a_1, \ldots, a_m \in k[X_1, \ldots, X_n]$ $(I = \langle f_1, \ldots, f_m \rangle)$.

Proof.   This follows from Theorem 5.4.7 and Corollary 5.4.5.   □

5.5   Newton revisited

Let us return to the question in the introduction to this chapter. Is there a systematic way of writing $X^4 + Y^4$ as a polynomial in $X + Y$ and XY? The answer is yes, and it is a nice consequence of the theory of Gröbner bases. In a more general setting we let $f, f_1, \ldots, f_r \in k[X_1, \ldots, X_n]$. We wish to decide whether the polynomial f can be written as $P(f_1, \ldots, f_r)$, where $P \in k[T_1, \ldots, T_r]$, and find P if this is the case. Consider the polynomial ring $A = k[X_1, \ldots, X_n, T_1, \ldots, T_r]$. If we can write

$$f = a_1(T_1 - f_1) + \cdots + a_r(T_r - f_r) + h, \qquad (5.5)$$

where $h \in k[T_1, \ldots, T_r]$ and $a_1, \ldots, a_r \in A$, then we may put $T_i = f_i$ so that $f = h(f_1, \ldots, f_r)$ and we can take $P = h$. Let $I \subseteq A$ be the ideal $\langle T_1 - f_1, \ldots, T_r - f_r \rangle$. If $f = P(f_1, \ldots, f_r)$, where $P \in k[T_1, \ldots, T_r]$, then (see Exercise 5.17)

$$f(X_1, \ldots, X_n) - P(T_1, \ldots, T_r) \in I. \qquad (5.6)$$

Therefore

$$f = a_1(T_1 - f_1) + \cdots + a_r(T_r - f_r) + P$$

for suitable $a_1, \ldots, a_r \in A$. How do we find the polynomial P? This is where the theory of Gröbner bases comes in handy. It gives the following surprising result.

Theorem 5.5.1   *Let $f, f_1, \ldots, f_r \in k[X_1, \ldots, X_n]$. Let I be the ideal*

$$I = \langle T_1 - f_1, \ldots, T_r - f_r \rangle$$

in the polynomial ring $A = k[X_1, \ldots, X_n, T_1, \ldots, T_r]$ and $\leq$ the lexicographic ordering given by

$$X_1 \geq \cdots \geq X_n \geq T_1 \geq \cdots \geq T_r.$$

Let G be a Gröbner basis of I with respect to $\leq$. Then f can be written as a polynomial in $f_1, \ldots, f_r$ if and only if

$$f^G \in k[T_1, \ldots, T_r].$$

In this case $f = f^G(f_1, \ldots, f_r)$.

Proof. Let $G = (g_1, \ldots, g_N)$ be a Gröbner basis for I with respect to $\leq$. Then the division algorithm gives

$$f = a'_1 g_1 + \cdots + a'_N g_N + f^G$$

for $a'_1, \ldots, a'_N \in A$. Since $\langle g_1, \ldots, g_N \rangle = I$, we can find $a_1, \ldots, a_r \in A$ such that

$$f = a_1(T_1 - f_1) + \cdots + a_r(T_r - f_r) + f^G.$$

If $f^G \in k[T_1, \ldots, T_r]$ then it follows that $f = f^G(f_1, \ldots, f_r)$ by (5.5).

If, however, there is a polynomial $P \in k[T_1, \ldots, T_r]$ such that $f = P(f_1, \ldots, f_r)$ then

$$f = a_1(T_1 - f_1) + \cdots + a_r(T_r - f_r) + P, \tag{5.7}$$

where $a_1, \ldots, a_r \in A$, by (5.6). We will prove that $f^G \in k[T_1, \ldots, T_r]$ in this case. This is done by running through the division algorithm with f and the Gröbner basis G. We may rewrite (5.7) as

$$f = b_1 g_1 + \cdots + b_N g_N + P$$

for suitable $b_1, \ldots, b_N \in A$. Notice that the invariant expression (5.2) of the division algorithm (Proposition 5.3.1) is satisfied by $s = P$ and $r = 0$ (using $b_1, \ldots, b_N$ as values for the coefficients of $g_1, \ldots, g_N$). If $\mathrm{in}_{\leq}(g_j)$ divides the $\mathrm{in}_{\leq}(s)$ entering (5.3) of the division algorithm (see the proof of Proposition 5.3.1), then $\mathrm{in}_{\leq}(g_j) \leq \mathrm{in}_{\leq}(s)$. This implies that $\mathrm{in}_{\leq}(g_j) \in k[T_1, \ldots, T_r]$. Therefore $g_j \in k[T_1, \ldots, T_r]$ if $s \in k[T_1, \ldots, T_r]$. Here it is important that the term ordering is lexicographic with $X_1 \geq \cdots \geq X_n \geq T_1 \geq \cdots \geq T_r$. So the assignment in (5.3) satisfies $s - (\mathrm{in}_{\leq}(s)/\mathrm{in}_{\leq}(g_j))g_j \in k[T_1, \ldots, T_r]$. Since we are

$$\underline{X^4} + Y^4 \quad : \quad (\underline{-X} - Y + T_1, \underline{Y^2} - YT_1 + T_2)$$
$$\underline{X^4 + X^3Y - X^3T_1}$$
$$\quad \underline{-X^3Y} + X^3T_1 + Y^4$$
$$\quad \underline{-X^3Y - X^2Y^2 + X^2YT_1}$$
$$\quad\quad \underline{X^3T_1} + X^2Y^2 - X^2YT_1 + Y^4$$
$$\quad\quad \underline{X^3T_1 + X^2YT_1 - X^2T_1^2}$$
$$\quad\quad\quad \underline{X^2Y^2} - 2X^2YT_1 + X^2T_1^2 + Y^4$$
$$\quad\quad\quad \underline{X^2Y^2 + XY^3 - XY^2T_1}$$
$$\quad\quad\quad\quad \underline{-2X^2YT_1} + X^2T_1^2 - XY^3 + XY^2T_1 + Y_4$$
$$\quad\quad\quad\quad \underline{-2X^2YT_1 - 2XY^2T_1 + 2XYT_1^2}$$
$$\quad\quad\quad\quad\quad \underline{X^2T_1^2} - XY^3 + 3XY^2T_1 - 2XYT_1^2 + Y^4$$
$$\quad\quad\quad\quad\quad \underline{X^2T_1^2 + XYT_1^2 - XT_1^3}$$
$$\quad\quad\quad\quad\quad\quad \underline{-XY^3} + 3XY^2T_1 - 3XYT_1^2 + XT_1^3 + Y^4$$
$$\quad\quad\quad\quad\quad\quad \underline{-XY^3 - Y^4 + Y^3T_1}$$
$$\quad\quad\quad\quad\quad\quad\quad \underline{3XY^2T_1} - 3XYT_1^2 + XT_1^3 + 2Y^4 - Y^3T_1$$
$$\quad\quad\quad\quad\quad\quad\quad \underline{3XY^2T_1 + 3Y^3T_1 - 3Y^2T_1^2}$$
$$\quad\quad\quad\quad\quad\quad\quad\quad \underline{-3XYT_1^2} + XT_1^3 + 2Y^4 - 4Y^3T_1 + 3Y^2T_1^2$$
$$\quad\quad\quad\quad\quad\quad\quad\quad \underline{-3XYT_1^2 - 3Y^2T_1^2 + 3YT_1^3}$$
$$\quad\quad\quad\quad\quad\quad\quad\quad\quad \underline{XT_1^3} + 2Y^4 - 4Y^3T_1 + 6Y^2T_1^2 - 3YT_1^3$$
$$\quad\quad\quad\quad\quad\quad\quad\quad\quad \underline{XT_1^3 + YT_1^3 - T_1^4}$$
$$\quad\quad\quad\quad\quad\quad\quad\quad\quad\quad \underline{2Y^4} - 4Y^3T_1 + 6Y^2T_1^2 - 4YT_1^3 + T_1^4$$
$$\quad\quad\quad\quad\quad\quad\quad\quad\quad\quad \underline{2Y^4 - 2Y^3T_1 + 2Y^2T_2}$$
$$\quad\quad\quad\quad\quad\quad\quad\quad\quad\quad\quad \underline{-2Y^3T_1} + 6Y^2T_1^2 - 2Y^2T_2 - 4YT_1^3 + T_1^4$$
$$\quad\quad\quad\quad\quad\quad\quad\quad\quad\quad\quad \underline{-2Y^3T_1 + 2Y^2T_1^2 - 2YT_1T_2}$$
$$\quad\quad\quad\quad\quad\quad\quad\quad\quad\quad\quad\quad \underline{4Y^2T_1^2} - 2Y^2T_2 - 4YT_1^3 + 2YT_1T_2 + T_1^4$$
$$\quad\quad\quad\quad\quad\quad\quad\quad\quad\quad\quad\quad \underline{4Y^2T_1^2 - 4YT_1^3 + 4T_1^2T_2}$$
$$\quad\quad\quad\quad\quad\quad\quad\quad\quad\quad\quad\quad\quad \underline{-2Y^2T_2} + 2YT_1T_2 + T_1^4 - 4T_1^2T_2$$
$$\quad\quad\quad\quad\quad\quad\quad\quad\quad\quad\quad\quad\quad \underline{-2Y^2T_2 + 2YT_1T_2 - 2T_2^2}$$
$$\quad\quad\quad\quad\quad\quad\quad\quad\quad\quad\quad\quad\quad\quad T_1^4 - 4T_1^2T_2 + 2T_2^2$$

Figure 5.1

moving terms from s to the remainder, in the division algorithm in (5.4), we will eventually end up with a remainder f^G in $k[T_1, \ldots, T_r]$.   □

Example 5.5.2 Let us return to the problem of writing $X^4 + Y^4$ as a polynomial in $X + Y$ and XY. Using Theorem 5.5.1 to address this we must find a Gröbner basis of $I = \langle T_1 - X - Y, T_2 - XY \rangle$ with respect to the lexicographic ordering $\leq$ given by $X \geq Y \geq T_1 \geq T_2$. You will see in the next section how to compute a Gröbner basis using Buchberger's algorithm. Let me reveal that a Gröbner basis for I with respect to $\leq$ is $G = (T_1 - X - Y, T_2 - T_1 Y + Y^2)$. Now we can use the division algorithm to find $(X^4 + Y^4)^G$. There are quite a number of steps, but (miraculously) we end with an expression involving only T_1 and T_2 as the remainder. Figure 5.1 shows the computation.

The computation in the figure shows that $(X^4 + Y^4)^G = T_1^4 - 4T_1^2 T_2 + 2T_2^2$. Without looking for clever algebraic tricks we have found a mechanical procedure. In this case the division algorithm shows that

$$X^4 + Y^4 = (X + Y)^4 - 4(X + Y)^2 XY + 2(XY)^2.$$

Notice that given any symmetric polynomial $f(X, Y)$ we can use the division algorithm to find $P = f^G$ such that $f = P(X + Y, XY)$. Theorem 5.5.1 is useful in that it gives a straightforward algorithm.

5.6   Buchberger's S-criterion

Theorem 5.4.7 shows the existence of a Gröbner basis for an ideal in a polynomial ring but gives no hint how to find it. There is a very nice (finite) criterion for a set of polynomials $F = (f_1, \ldots, f_m)$ to be a Gröbner basis. To a pair of polynomials f, g we associate the S-polynomial $S(f, g)$, which depends on the term ordering $\leq$. The S-polynomial $S(f, g)$ cancels initial terms in f and g according to the term ordering $\leq$. For example, $S(X^2 + Y, YX + 1) = Y(X^2 + Y) - X(YX + 1) = Y^2 - X$, where $\leq$ is the lexicographic ordering with $Y \leq X$. Buchberger's S-criterion says that F is a Gröbner basis for I if and only if $S(f_i, f_j)^F = 0$ for $1 \leq i < j \leq m$.

This turns out to be very useful in practice. It is also the basis of Buchberger's algorithm for finding Gröbner bases. If an S-polynomial S does not give a remainder S^F equal to zero then you simply add the remainder S^F to the list of polynomials and use Buchberger's S-criterion on this new list. This will eventually terminate (Buchberger's criterion will succeed).

A word of advice: no complicated or abstract mathematics is involved, just (very) clever calculations with polynomials. As a first approach to understanding Buchberger's algorithm you can go straight to subsection 5.6.2 after reading the statement of Theorem 5.6.8 and understanding the definition of S-polynomials (Definition 5.6.5). In the following, a term ordering $\leq$ is fixed on $R = k[X_1, \ldots, X_n]$.

5.6.1 The S-polynomials

Suppose we wish to check whether $(f_1, \ldots, f_m) \subseteq R \setminus \{0\}$ is a Gröbner basis. Let

$$f = a_1 f_1 + \cdots + a_m f_m \in \langle f_1, \ldots, f_m \rangle,$$

where $a_1, \ldots, a_m \in R$. Does $\text{in}_{\leq}(f_i)$ divide $\text{in}_{\leq}(f)$ for some $i = 1, \ldots, m$? Put $aX^v = \text{in}_{\leq}(f)$, $c_i X^{u_i} = \text{in}_{\leq}(a_i)$ and $d_i X^{v_i} = \text{in}_{\leq}(f_i)$ for $i = 1, \ldots, m$. Now introduce

$$\delta = \max {}_{\leq}\{v_i + u_i \mid i = 1, \ldots m\}.$$

Then it is impossible that $v \gneq v_i + u_i$ for every $i = 1, \ldots, m$, since the initial term of f has to be a k-linear combination of the initial terms $\text{in}_{\leq}(a_i f_i)$ for $a_i f_i \neq 0$. Therefore $v \leq \delta$. If $\delta = v$, we may assume that $\delta = v_1 + u_1 = \cdots = v_r + u_r$, where $r \leq m$ and $a_i f_i \neq 0$ for $i = 1, \ldots, r$. Then

$$aX^v = (c_1 d_1 + \cdots + c_r d_r)X^{u_1 + v_1}.$$

In this case $d_1 X^{v_1} = \text{in}_{\leq}(f_1)$ divides $aX^v = \text{in}_{\leq}(f)$. However, if $v < \delta$ there is cancellation of maximal terms on the right hand side, and $\text{in}_{\leq}(f)$ is not necessarily divisible by $\text{in}_{\leq}(f_i)$, for $i = 1, \ldots, m$. This is illustrated by the following example.

Example 5.6.1 Let $\leq$ be the lexicographic ordering given by $X \geq Y$, $I = \langle X^2 + Y, X^2 Y + 1 \rangle \subseteq k[X, Y]$ and $f = Y^2 - 1 = Y(X^2 + Y) - (X^2 Y + 1) \in I$. Then $\text{in}_{\leq}(f) = Y^2$ but $X^2 \nmid Y^2$ and $X^2 Y \nmid Y^2$.

Our discussion leads to the following definition and proposition.

Definition 5.6.2 We say that $f \in R$ reduces to zero modulo $F = (f_1, \ldots f_m) \subseteq R \setminus \{0\}$ if there exist $a_1, \ldots, a_m \in R$ such that

$$f = a_1 f_1 + \cdots + a_m f_m \tag{5.8}$$

and $in_\le(a_i f_i) \le in_\le(f)$ if $a_i f_i \ne 0$. This is denoted

$$f \to_F 0.$$

Remark 5.6.3   Observe that $f \to_F 0$ if and only if the maximal initial terms in the summands on the right hand side of (5.8) do not cancel. Notice also that $f \to_F 0$ if $f^F = 0$. This is the last part of Proposition 5.3.1. However, one may have $f \to_F 0$ even though $f^F \ne 0$ (see Exercise 5.18).

Proposition 5.6.4   *Let $F = (f_1, \ldots, f_m)$ and $I = \langle f_1, \ldots, f_m \rangle$. If $f \to_F 0$ for every $f \in I$ then F is a Gröbner basis for I. If F is a Gröbner basis for I then $f^F = 0$ if and only if $f \to_F 0$ for $f \in I$.*

Proof.   Let $f \in I \setminus \{0\}$. The discussion at the beginning of this subsection shows that if $f \to_F 0$ then $in_\le(f)$ is divisible by $in_\le(f_j)$ for some $f_j \in F$. So if $f \to_F 0$ for every $f \in I$ it follows that F is a Gröbner basis for I. We have seen that $f^F = 0$ implies that $f \to_F 0$ by the last part of Proposition 5.3.1. If F is a Gröbner basis for I and $f \to_F 0$ then $f^F = 0$ since $f \in I$ (this is Proposition 5.4.2). $\square$

This is really not a useful test for a Gröbner basis. We need to check that every $f \in I$ reduces to zero. Using some clever manipulations one may find finitely many polynomials $S_1, \ldots, S_N \in I$ such that F is a Gröbner basis if and only if $S_i \to_F 0$ for $i = 1, \ldots, N$. We can in fact replace $S_i \to_F 0$ with $S_i^F = 0$, by Proposition 5.6.4. In this way we have an effective criterion for a Gröbner basis via the division algorithm provided that we can find $S_1, \ldots, S_N$. Let us see how to do this. Suppose that

$$f = a_1 f_1 + \cdots + a_m f_m \in I,$$

where $a_1, \ldots, a_m \in R$. Use the notation from the beginning of this subsection. Then

$$f = C + (a_1 - in_\le(a_1))f_1 + \cdots + (a_r - in_\le(a_r))f_r + a_{r+1}f_{r+1} + \cdots + a_m f_m,$$

where $C = in_\le(a_1)f_1 + \cdots + in_\le(a_r)f_r$. One crucial point to notice is that f is the sum of C and certain polynomials all of whose initial terms are $\preceq \delta$. If on the one hand $c_1 d_1 + \cdots + c_r d_r \ne 0$ then no cancellation among the initial terms occurs and $in_\le(f)$ is divisible by $in_\le(f_i)$ for some $i = 1, \ldots, m$, as we have already seen.

Assume on the other hand that $c_1 d_1 + \cdots + c_r d_r = 0$ (cancellation occurs among the initial terms). Put $g_i = X^{u_i} f_i / d_i$ and watch the following nice computational trick evolve:

$$C = c_1 d_1 g_1 + \cdots + c_r d_r g_r$$
$$= c_1 d_1 (g_1 - g_2) + (c_1 d_1 + c_2 d_2)(g_2 - g_3) + (c_1 d_1 + c_2 d_2 + c_3 d_3)(g_3 - g_4)$$
$$+ \cdots + (c_1 d_1 + \cdots + c_{r-1} d_{r-1})(g_{r-1} - g_r) + (c_1 d_1 + \cdots + c_r d_r) g_r.$$

This shows that C is a linear combination of $g_i - g_j = X^{u_i} f_i / d_i - X^{u_j} f_j / d_j$. From this we get the crucial S-polynomials. Observe that $u_i + v_i = u_j + v_j$ as vectors in $\mathbb{N}^n$ (the initial terms of g_i and g_j cancel). Now define $w_{ij} \in \mathbb{N}^n$ by $X^{w_{ij}} = \mathrm{lcm}(X^{v_i}, X^{v_j})$. Then

$$g_i - g_j = \frac{X^{u_i} f_i}{d_i} - \frac{X^{v_j} f_j}{d_j}$$
$$= X^{\zeta} \left(\frac{X^{w_{ij}}}{d_i X^{v_i}} f_i - \frac{X^{w_{ij}}}{d_j X^{v_j}} f_j \right),$$

where $\zeta + w_{ij} = u_i + v_i = u_j + v_j$. Notice the cancellation of the two initial terms in

$$\frac{X^{w_{ij}}}{d_i X^{v_i}} f_i - \frac{X^{w_{ij}}}{d_j X^{v_j}} f_j.$$

This naturally leads us to the following definition.

Definition 5.6.5   The S-polynomial of two non-zero polynomials f and g with respect to a term ordering $\leq$ is defined as

$$S(f, g) = \frac{X^w}{\mathrm{in}_{\leq}(f)} f - \frac{X^w}{\mathrm{in}_{\leq}(g)} g,$$

where X^w is a least common multiple of $\mathrm{in}_{\leq}(f)$ and $\mathrm{in}_{\leq}(g)$.

The formal definition of S-polynomials may take some time to digest. Intuitively one just multiplies the initial terms of f and g up to a least common multiple. The letter S in S-polynomial stands for "syzygy." This is a concept from Hilbert's theory of syzygies for polynomial rings. A syzygy is a term from astronomy. It refers to a straight-line configuration of three celestial bodies. The moon is in syzygy with the Earth and the Sun when it is new or full.

Example 5.6.6 Let $\leq$ be the lexicographic ordering given by $X \geq Y$ in $k[X, Y]$. Then $\text{lcm}(X^2, X^2Y) = X^2Y$, and

$$
\begin{aligned}
S(X^2 + Y, \ X^2Y + 1) &= \frac{X^2Y}{X^2}(X^2 + Y) - \frac{X^2Y}{X^2Y}(X^2Y + 1) \\
&= Y(X^2 + Y) - (X^2Y + 1) \\
&= Y^2 - 1.
\end{aligned}
$$

We have shown that

$$
\begin{aligned}
C &= \text{in}_{\leq}(a_1)f_1 + \cdots + \text{in}_{\leq}(a_r)f_r \\
&= b_1 X^{\zeta_1} S(f_1, f_2) + \cdots + b_{r-1} X^{\zeta_{r-1}} S(f_{r-1}, f_r) \quad (5.9)
\end{aligned}
$$

with $b_i \in k$ and $\text{in}_{\leq}(X^{\zeta_i} S(f_i, f_{i+1})) \lneq \delta$. This calculation is crucial for the proof of the following important insight.

Lemma 5.6.7 Let $F = (f_1, \ldots, f_m)$ and $I = \langle f_1, \ldots, f_m \rangle$. If $S(f_i, f_j) \to_F 0$ for every $i, j = 1, \ldots, m$ then $f \to_F 0$ for every $f \in I$.

Proof. Let $f = a_1 f_1 + \cdots + a_m f_m \in I$, where $a_1, \ldots, a_m \in R$. Since $S(f_i, f_j) \to_F 0$, we have

$$
S(f_i, f_j) = e_1 f_1 + \cdots + e_m f_m
$$

for $e_1, \ldots, e_m \in R$, where $\text{in}_{\leq}(e_l f_l) \leq \text{in}_{\leq}(S(f_i, f_j))$ for $l = 1, \ldots, m$. Recall that

$$
\begin{aligned}
f &= C + (a_1 - \text{in}_{\leq}(a_1))f_1 + \ldots + (a_r - \text{in}_{\leq}(a_r))f_r + a_{r+1}f_{r+1} + \ldots \\
&\quad + a_m f_m,
\end{aligned}
$$

where $C = \text{in}_{\leq}(a_1)f_1 + \cdots + \text{in}_{\leq}(a_r)f_r$ and $\text{in}_{\leq}(a_1 f_1), \ldots, \text{in}(a_r f_r)$ are the maximal initial terms in the summands $a_1 f_1, \ldots, a_m f_m$. Now insert the expression for $S(f_i, f_j)$ into (5.9) to get

$$
f = h_1 f_1 + \cdots + h_m f_m
$$

with $\max\{\text{in}_{\leq}(h_i f_i) \mid h_i f_i \neq 0, \ i = 1, \ldots, n\} \lneq \delta$. This means that *if the maximal initial terms on the right hand side of an expression $f = a_1 f_1 + \cdots + a_n f_n$ cancel and $S(f_i, f_j) \to_F 0$ then there is another expression $f = h_1 f_1 + \cdots + h_n f_n$ for which the maximal initial term in the summands on the right hand side is strictly less than the maximal initial term in the first expression.* By Lemma A.3.8 we will eventually end up with an expression

$$
f = b_1 f_1 + \cdots + b_m f_m,
$$

where the maximal initial term δ in the summands on the right hand side is $\text{in}_{\leq}(f)$. This means that $f \rightarrow_F 0$. $\qquad\square$

5.6.2 The S-criterion

Theorem 5.6.8 (Buchberger) *A sequence $F = (f_1, \ldots, f_m)$ of polynomials is a Gröbner basis if and only if $S(f_i, f_j) \rightarrow_F 0$ for $1 \leq i < j \leq m$.*

Proof. This is a consequence of Proposition 5.6.4 and Lemma 5.6.7. $\qquad\square$

Corollary 5.6.9 *A sequence $F = (f_1, \ldots, f_m)$ of polynomials is a Gröbner basis if and only if $S(f_i, f_j)^F = 0$ for $1 \leq i < j \leq m$.*

Proof. If $S(f_i, f_j)^F = 0$ for $1 \leq i < j \leq m$ then $S(f_i, f_j) \rightarrow_F 0$ for $1 \leq i < j \leq m$ and F is a Gröbner basis by Theorem 5.6.8. Conversely, if F is a Gröbner basis then $S(f_i, f_j)^F = 0$ by Proposition 5.4.2, since $S(f_i, f_j) \in \langle f_1, \ldots, f_m \rangle$. $\qquad\square$

5.7 Buchberger's algorithm

The Buchberger S-criterion (Corollary 5.6.9) is a systematic way of testing whether a set of polynomials $F = (f_1, \ldots, f_m)$ is a Gröbner basis. Compute the remainders of the S-polynomials $S(f_i, f_j)$, where $1 \leq i < j \leq m$. On the one hand, if they are all zero then F is a Gröbner basis. On the other hand, if one $S(f_i, f_j)^F \neq 0$ then we simply add it to F to obtain a new list

$$F' = F \cup \{S(f_i, f_j)^F\} = (f_1, \ldots, f_m, S(f_i, f_j)^F),$$

hoping that F' will turn out to be a Gröbner basis for $I = \langle f_1, \ldots, f_m \rangle$. Notice that F' and F generate the same ideal since $S(f_i, f_j)^F \in I$.

We can continue adding remainders of S-polynomials to our list. This is a somewhat daring step. We have no guarantee that this procedure will ever stop. Let us try it out on an example.

Example 5.7.1 Suppose we have the lexicographic ordering given by $X \geq Y$ on $k[X, Y]$ and $F = (X^2 + Y, X^2Y + 1)$. Then $S(X^2 + Y, X^2Y + 1) = Y^2 - 1$. This also becomes the remainder in the division algorithm, since none of the terms Y^2 and -1 is divisible by $\text{in}_{\leq}(X^2 + Y) = X^2$ or $\text{in}_{\leq}(X^2Y + 1) = X^2Y$.

Thus

$$S(X^2 + Y, \ X^2 Y + 1)^F = Y^2 - 1.$$

Now let

$$F' = F \cup \{Y^2 - 1\} = (X^2 + Y, \ X^2 Y + 1, \ Y^2 - 1).$$

To check whether this is a Gröbner basis, we have to compute $S(X^2 + Y, \ Y^2 - 1)^{F'}$ and $S(X^2 Y + 1, \ Y^2 - 1)^{F'}$ and see whether they are zero. It is not necessary to compute $S(X^2 + Y, \ X^2 Y + 1)^{F'}$, as this is zero because $S(X^2 + Y, \ X^2 Y + 1) = 1 \cdot (Y^2 - 1) + 0$. Now

$$S(X^2 + Y, \ Y^2 - 1) = Y^2(X^2 + Y) - X^2(Y^2 - 1) = Y^3 + X^2.$$

The division algorithm gives $Y^3 + X^2 = 1 \cdot (X^2 + Y) + Y \cdot (Y^2 - 1)$, so the remainder is zero. Finally

$$S(X^2 Y + 1, \ Y^2 - 1) = Y(X^2 Y + 1) - X^2(Y^2 - 1) = 1 \cdot (X^2 + Y),$$

which also has zero remainder. By Corollary 5.6.9,

$$(X^2 + Y, \ X^2 Y + 1, \ Y^2 - 1)$$

is a Gröbner basis.

The process of continuously adding non-zero remainders of S-polynomials is called *Buchberger's algorithm*. There are numerous ways of implementing it. The workhorse in the algorithm is the division algorithm and one usually wants as few divisions as possible. We will not go into implementation details but simply prove that the algorithm terminates.

Theorem 5.7.2 *Buchberger's algorithm terminates and the output is a Gröbner basis.*

Proof. Let $F = (f_1, \ldots, f_s)$ be the list of polynomials in a step of Buchberger's algorithm. Suppose that $1 \le i < j \le s$ and $S^F \ne 0$, where $S = S(f_i, f_j)$. Since S^F is a remainder coming from the division algorithm with $F = (f_1, \ldots, f_s)$, no term in S^F is divisible by any of $\mathrm{in}_{\le}(f_1), \ldots, \mathrm{in}_{\le}(f_s)$. So we may prove that the algorithm terminates by proving that for any sequence of terms $T = (t_1, t_2, \ldots)$ there exists a number $N \in \mathbb{N}$ such that if $i \ge N$ then t_i is divisible by t_j, where $j < N$. Dickson's lemma (Lemma 5.1.5) implies that there are finitely many terms $t_{i_1}, \ldots, t_{i_r} \in T$

such that every term $t \in T$ is divisible by one of $t_{i_1}, \ldots, t_{i_r}$. Putting $N = \max(i_1, \ldots, i_r)$ we get the result. $\qquad\square$

The following lemma sometimes simplifies the computations in Buchberger's algorithm considerably.

Lemma 5.7.3 *Let $\leq$ be a term ordering on $R = k[X_1, \ldots, X_n]$. Let $f, g \in R$ and suppose that $\mathrm{in}_\leq(f)$ and $\mathrm{in}_\leq(g)$ have no common divisors (except constants). Then*

$$S(f, g) \to_{(f,g)} 0.$$

Proof. Put $r = f - \mathrm{in}_\leq(f)$ and $s = g - \mathrm{in}_\leq(g)$. Then

$$S(f, g) = (g - s)f - (f - r)g = rg - sf.$$

If the initial terms in rg and sf cancel then

$$\mathrm{in}_\leq(r)\,\mathrm{in}_\leq(g) = \mathrm{in}_\leq(s)\,\mathrm{in}_\leq(f).$$

This implies that $\mathrm{in}_\leq(f) \mid \mathrm{in}_\leq(r)$, contradicting that $\mathrm{in}_\leq(r) < \mathrm{in}_\leq(f)$. So the initial terms of rg and sf do not cancel. This shows that $S(f, g) \to_{(f,g)} 0$. $\qquad\square$

Example 5.7.4 Let $F = (T_1 - X - Y, T_2 - XY) \subseteq k[X, Y, T_1, T_2]$. Then F is already a Gröbner basis with respect to the lexicographic term ordering given by $T_1 \geq T_2 \geq X \geq Y$. This is a consequence of Theorem 5.6.8 and Lemma 5.7.3. However, if the term ordering is given by $X \geq Y \geq T_1 \geq T_2$, as in Example 5.5.2, then

$$S = S(T_1 - X - Y, T_2 - XY) = Y(T_1 - X - Y)$$
$$- (T_2 - XY) = YT_1 - Y^2 - T_2.$$

Notice that $S^F = S$. Using Corollary 5.6.9 you should check that $F \cup \{S\}$ is a Gröbner basis.

Example 5.7.5 Looking innocent at first, Gröbner bases can be hairy beasts that are extremely time consuming to compute and very dependent on the term ordering. Take for example ([23], Example 3.9) the ideal

$$I = \langle x^5 + y^3 + z^2 - 1, \ x^2 + y^2 + z - 1, \ x^6 + y^5 + z^3 - 1 \rangle$$

in $\mathbb{Q}[x, y, z]$. A Gröbner basis of I with respect to the lexicographic ordering $z \geq y \geq x$ is the monstrous list of polynomials seen in Figure 5.2.

$(225\,x^4 + 675\,x^5 + 705\,x^6 + 315\,x^7 + 100\,x^8 - 555\,x^9 - 1946\,x^{10} -$

$1983\,x^{11} - 10\,x^{12} + 1225\,x^{13} + 697\,x^{14} + 195\,x^{15} + 226\,x^{16} +$

$139\,x^{17} - x^{18} - 13\,x^{19} + 3\,x^{20} + 2\,x^{21} + x^{22}, 4794799513743465\,x^4 +$

$9461645755921935\,x^5 + 5609230341167770\,x^6 + 1305539383606500\,x^7 +$

$426289252230518\,x^8 - 12718603398056543\,x^9 - 28161279400718496\,x^{10} -$

$13641002940967260\,x^{11} + 13303041747347884\,x^{12} + 12841472514397999\,x^{13} +$

$1936021990228677\,x^{14} + 2115618449641410\,x^{15} + 2686197967416241\,x^{16} +$

$266417434391307\,x^{17} - 308399336177560\,x^{18} + 40028515719740\,x^{19} +$

$22083510506531\,x^{20} + 20898699599882\,x^{21} - 307985585745030\,x^4\,y +$

$307985585745030\,x^5\,y, 37955678888811405\,x^4 + 40874650161525720\,x^5 -$

$3971051857805515\,x^6 + 8461551779562300\,x^7 - 7477091544441736\,x^8 -$

$133100833227195819\,x^9 - 130427012317955273\,x^{10} + 96308769549551000\,x^{11} +$

$112430217894147542\,x^{12} - 28978302929820573\,x^{13} - 8147851966720744\,x^{14} +$

$23240432665880855\,x^{15} - 2547153248711687\,x^{16} - 6558796078633904\,x^{17} +$

$1957860431279775\,x^{18} - 154503618530810\,x^{19} + 226403721396233\,x^{20} -$

$92968302338769\,x^{21} + 9239567572350900\,x^2\,y - 9239567572350900\,x^3\,y -$

$9239567572350900\,x^2\,y^2 + 9239567572350900\,x^3\,y^2, -92395675723509000\,x^2 +$

$267932368916755545\,x^4 + 607600416419937750\,x^5 + 326949813554222075\,x^6 -$

$32115739051910620\,x^7 - 858543129560584\,x^8 - 533880675743739115\,x^9 -$

$1553067597584776499\,x^{10} - 1058691906621826800\,x^{11} + 691613184599027638\,x^{12} +$

$932606563955672291\,x^{13} + 15138939075195079 4\,x^{14} + 95707520810719369\,x^{15} +$

$185431646079855213\,x^{16} + 30397871204445410\,x^{17} - 24246152848015907\,x^{18} +$

$2994483268700962\,x^{19} + 1053727522296225\,x^{20} + 1579303619755253\,x^{21} -$

$92395675723509000\,y^2 + 92395675723509000\,x^2\,y^2 + 92395675723509000\,y^3,$

$- 1 + x^2 + y^2 + z).$

Figure 5.2

Surprisingly, there is a term ordering $\leq$ such that the Gröbner basis of I with respect to $\leq$ is (see Exercise 5.29)

$$(x^5 + y^3 + z^2 - 1, \; x^2 + y^2 + z - 1, \; x^6 + y^5 + z^3 - 1).$$

Here Lemma 5.7.3 is very useful.

5.8 The reduced Gröbner basis

In the following, we work with a fixed term ordering $\leq$ in $R = k[X_1, \ldots, X_n]$. A Gröbner basis $(f_1, \ldots, f_m)$ for an ideal $I \subseteq R$ is not unique. You can always add another polynomial $f \in I$ to the list $(f_1, \ldots, f_n)$ and it will still be a Gröbner basis (see Exercise 5.15). We need a more well behaved object that is unique. We may begin by observing that if we have a Gröbner basis $(f_1, \ldots, f_m)$ for the ideal I and $\mathrm{in}_\leq(f_1)$ is divisible by one of $\mathrm{in}_\leq(f_2), \ldots, \mathrm{in}_\leq(f_m)$ then $(f_2, \ldots, f_m)$ is a smaller Gröbner basis for I. Assume that $\mathrm{in}_\leq(f_i) \mid \mathrm{in}_\leq(f_1)$; then $\mathrm{in}_\leq(f_i) \mid \mathrm{in}_\leq(f)$ if $\mathrm{in}_\leq(f_1) \mid \mathrm{in}_\leq(f)$, where $f \in I$. So $(f_2, \ldots, f_m)$ is a Gröbner basis for I and $I = \langle f_2, \ldots, f_m \rangle$ by Corollary 5.4.5. This shows that an efficient strategy for cutting down on the size of a Gröbner basis is to throw away generators f whose initial term $\mathrm{in}_\leq(f)$ is divisible by the initial term of one of the other generators. This leads to the definition of a minimal Gröbner basis.

Definition 5.8.1   A *minimal Gröbner basis* $(f_1, \ldots, f_m)$ is a Gröbner basis such that

(i) $\mathrm{in}_\leq(f_i)$ is not divisible by $\mathrm{in}_\leq(f_j)$ for $i \neq j$,
(ii) the coefficient of $\mathrm{in}_\leq(f_i)$ is 1.

A minimal Gröbner basis is still not unique even though it has the minimal number of elements! The unique object is the reduced Gröbner basis.

Definition 5.8.2   A *reduced Gröbner basis* $(f_1, \ldots, f_m)$ is a minimal Gröbner basis such that no term (not just the initial term) in f_i is divisible by $\mathrm{in}_\leq(f_j)$ for $i \neq j$.

Theorem 5.8.3   *Every ideal $I \subseteq k[X_1, \ldots, X_n]$ has a unique reduced Gröbner basis.*

Proof. If $(f_1, \ldots, f_m)$ and $(g_1, \ldots, g_{m'})$ are two reduced Gröbner bases of I, we must have $m = m'$ and

$$\mathrm{in}_\le(f_1) = \mathrm{in}_\le(g_1),$$

$$\vdots$$

$$\mathrm{in}_\le(f_m) = \mathrm{in}_\le(g_m),$$

rearranging $g_1, \ldots, g_m$ if necessary. Here is why. We know that some $\mathrm{in}_\le(f_j)$ divides $\mathrm{in}_\le(g_1)$. We may assume by rearranging that $j = 1$. We also know that some $\mathrm{in}_\le(g_i)$ divides $\mathrm{in}_\le(f_1)$. Here $i = 1$, because $\mathrm{in}_\le(g_1)$ is divisible by $\mathrm{in}_\le(g_i)$. This gives that $\mathrm{in}_\le(f_1) = \mathrm{in}_\le(g_1)$, since the coefficient in both is 1. The same argument applies to the other generators, and we end up with $m = m'$ identical initial terms.

Now we wish to prove that $f_1 = g_1, \ldots, f_n = g_n$ in order to prove the uniqueness of the reduced Gröbner basis. Consider $f_1 - g_1$. The initial terms in f_1 and g_1 cancel. By definition of a reduced Gröbner basis none of the terms in $f_1 - g_1$ is divisible by any $\mathrm{in}_\le(f_1), \ldots, \mathrm{in}_\le(f_n)$ (here we include $\mathrm{in}_\le(f_1)$ because it has been canceled already in $f_1 - g_1$). This means that $f_1 - g_1$ is the remainder after division by $f_1, \ldots, f_n$. But since $f_1 - g_1 \in I$ we must have $f_1 - g_1 = 0$ by Proposition 5.4.2. The same procedure applies to the other generators.

Every ideal has a minimal Gröbner basis $(f_1, \ldots, f_m)$ by the reasoning at the beginning of Section 5.8. The existence of a reduced Gröbner basis is deduced as follows: replace f_1 by the remainder of f_1 divided by $f_2, \ldots, f_m$. With this new f_1, replace f_2 by the remainder of f_2 divided by $f_1, f_3, \ldots, f_n$. Continue this procedure until f_m is replaced by its remainder divided by $f_1, \ldots, f_{m-1}$. *Notice* that the initial terms of the original $f_1, \ldots, f_m$ will survive and that we still have a Gröbner basis. In the end no term of f_i is divisible by $\mathrm{in}_\le(f_j)$ for $i \ne j$. Thus we end up with a reduced Gröbner basis. $\square$

Example 5.8.4 In Example 5.7.1 we saw that $(X^2 + Y, X^2Y + 1, Y^2 - 1)$ is a Gröbner basis for the ideal $I = \langle X^2 + Y, X^2Y + 1 \rangle$ with respect to the lexicographic ordering $\le$, where $Y \le X$. It is not minimal, though! The second generator has initial term X^2Y, which is divisible by the initial term X^2 of the first generator. We can thus leave out the middle generator, ending up with

$$(X^2 + Y, Y^2 - 1)$$

which in fact is *the* reduced Gröbner basis of I for the term ordering $\le$.

Example 5.8.5   The Gröbner basis

$$G = (T_1 - X - Y, \ T_2 - XY, \ YT_1 - Y^2 - T_2) \subseteq k[X, Y, T_1, T_2]$$

from Example 5.7.4 is not minimal. The reason is that $\mathrm{in}_{\le}(T_2 - XY) = -XY$ is divisible by $\mathrm{in}_{\le}(T_1 - X - Y) = -X$. Leaving out the middle generator we get the Gröbner basis

$$G' = (T_1 - X - Y, \ YT_1 - Y^2 - T_2).$$

This is the Gröbner basis used in Example 5.5.2. You may check that G' is the reduced Gröbner basis when multiplied by -1.

5.9 Solving equations using Gröbner bases

Suppose we are given a set of polynomial equations in n variables over a field k:

$$f_1(x_1, \ldots, x_n) = 0,$$
$$f_2(x_1, \ldots, x_n) = 0,$$
$$\vdots$$
$$f_m(x_1, \ldots, x_n) = 0.$$

Just as in the old days of algebra, we want to find the solutions of these equations. If $n = 1$ we have a system of polynomial equations in just one variable x_1. This can be solved using the Euclidean algorithm: we know that the ideal $\langle f_1, \ldots, f_m \rangle \subseteq k[x_1]$ generated by $f_1, \ldots, f_m \in k[x_1]$ is a principal ideal $\langle f \rangle$, generated by a greatest common divisor f of $f_1, \ldots, f_n$. It follows that $f_1(x) = \cdots = f_m(x) = 0$ if and only if $f(x) = 0$. So we have reduced to the case of just one equation. Let $V(f_1, \ldots, f_m)$ denote

$$\{(a_1, \ldots, a_n) \in k^n \mid f_i(a_1, \ldots, a_n) = 0 \text{ for every } i = 1, \ldots, m\},$$

the set of solutions of the system of equations. Then $V(f_1, \ldots, f_m)$ is also given by

$$V(I) = \{(a_1, \ldots, a_n) \in k^n \mid f(a_1, \ldots, a_n) = 0 \text{ for every } f \in I\},$$

where I denotes the ideal generated by $f_1, \ldots, f_m$ (see Exercise 5.31). The ideal I represents all the equations we can get by "combining" $f_1, \ldots, f_m$. In particular, if we have a Gröbner basis $(g_1, \ldots, g_r)$ of I we get

$$V(f_1, \ldots, f_m) = V(g_1, \ldots, g_r).$$

The point is that the equations

$$g_1(x_1, \ldots, x_n) = 0,$$
$$g_2(x_1, \ldots, x_n) = 0,$$
$$\vdots$$
$$g_r(x_1, \ldots, x_n) = 0$$

are often much easier to solve.

This is the basis for doing "Gaussian" elimination on our system of equations using Gröbner bases. We wish to eliminate variables by combining some equations to get equations with fewer variables. The ideal situation is if the system of equations consists of some equations containing the variables $x_1, \ldots, x_n$, some equations containing the variables $x_2, \ldots, x_n$ and ... and some equations containing only x_n. Then we could begin by solving the equations involving only x_n, insert our solutions into the equations involving only x_{n-1} and x_n and so forth. Thereby we only have to solve equations involving one variable. The process of eliminating variables can be formulated as that of finding polynomials in I involving only x_1, polynomials in I involving only x_1, x_2 and so on. Viewing I as the equations that we can deduce by combining $f_1, \ldots, f_m$ we wish to find

$$I \cap k[x_1],$$
$$I \cap k[x_1, x_2],$$
$$\vdots$$
$$I \cap k[x_1, \ldots, x_{n-1}].$$

The following theorem is almost too good to be true.

Theorem 5.9.1 *Let G be a Gröbner basis for an ideal $I \subseteq k[X_1, \ldots, X_n]$ with respect to the lexicographic ordering $\leq$ given by $X_1 \leq X_2 \leq \cdots \leq X_n$. Then $G \cap k[X_1, \ldots, X_i]$ is a Gröbner basis for the ideal $I \cap k[X_1, \ldots, X_i]$ in $k[X_1, \ldots, X_i]$ with respect to the lexicographic ordering $\leq$ for the polynomials in $k[X_1, \ldots, X_i]$.*

Let $G' = G \cap k[X_1, \ldots, X_i]$. Suppose that $f \in I \cap k[X_1, \ldots, X_i]$. Then $\text{in}_{\leq}(g) \mid \text{in}_{\leq}(f)$ for some $g \in G$ using Definition 5.4.1. On the other hand

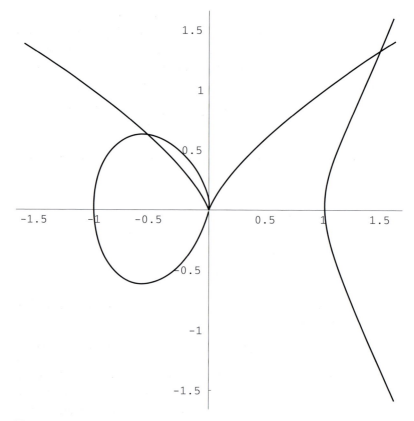

Figure 5.3

the terms in g are all smaller than $\mathrm{in}_{\le}(g)$ in our lexicographic term ordering. This tells us (why??) that $g \in G'$. Therefore G' is a Gröbner basis for $I \cap k[X_1, \ldots, X_i]$ with respect to $\le$ for the polynomials in $k[X_1, \ldots, X_i]$.

Example 5.9.2 Let us find the solutions to the system of equations

$$Y^2 - X^3 + X = 0,$$
$$Y^3 - X^2 = 0 \tag{5.10}$$

in $\mathbb{R}^2$. This corresponds to finding the points of intersection between the curves shown in Figure 5.3.

To solve (5.10) we need to transform it to another system of equations according to Theorem 5.9.1. We will do this by computing a Gröbner basis for $\langle Y^2 - X^3 + X, Y^3 - X^2 \rangle$ with respect to the lexicographic ordering $\leq$ where $X \geq Y$. A straightforward application of Buchbergers algorithm (even though the algorithm needs a few steps here) gives the Gröbner basis

$$(Y^2 - \underline{X}^3 + X, \, Y^3 - \underline{X}^2, \, -X - Y^2 + \underline{XY}^3, \, \underline{XY}^2 + Y^3 - Y^6,$$
$$Y^3 - Y^4 - 2Y^6 + \underline{Y}^9, \, -\underline{X} - Y^2 - Y^4 + Y^7),$$

where the initial terms are underlined. From this we see that the reduced Gröbner basis is

$$(Y^3 - Y^4 - 2Y^6 + Y^9, \, X + Y^2 + Y^4 - Y^7).$$

So finding the solutions to (5.10) is equivalent to solving

$$Y^3 - Y^4 - 2Y^6 + Y^9 = 0,$$
$$X + Y^2 + Y^4 - Y^7 = 0.$$

This is much more manageable than solving the original system (5.10). Now we can find the solutions to the equation

$$Y^3 - Y^4 - 2Y^6 + Y^9 = Y^3(1 - Y - 2Y^3 + Y^6) = 0 \qquad (5.11)$$

and plug them into $X + Y^2 + Y^4 - Y^7$ and get the corresponding X-values. Using numerical approximations (and a computer) one finds apart from $Y = 0$ that $Y = 0.605423$ and $Y = 1.2876$ are approximate real solutions to (5.11). So the real solutions to (5.10) are $(0, 0)$, $(-0.471073, 0.605423)$ and $(1.46109, 1.2876)$.

Notice that $\mathbb{R}[Y] \cap \langle Y^2 - X^3 + X, Y^3 - X^2 \rangle = \langle Y^3 - Y^4 - 2Y^6 + Y^9 \rangle$ by Theorem 5.9.1.

It is worth pointing out that all the clever algebraic tricks one might come up with solving a system of polynomial equations have been translated into a precise method using Gröbner bases.

5.10 Exercises

1. In Section 5.1 the set $R[\mathbb{N}^n]$ was introduced along with an addition and a multiplication. Let $f, g, h \in R[\mathbb{N}^n]$.
   (i) Prove that $f + g, fg \in R[\mathbb{N}^n]$.
   (ii) Prove that $fg = gf$.

(iii) Prove that $f(g + h) = fg + fh$.

(iv) Prove that $f(gh) = (fg)h$ by reducing to the case $h = cX^v$.

2. Prove that the ideal $\langle X, Y \rangle \subseteq \mathbb{Q}[X, Y]$ is not a principal ideal, by assuming that there exists $f \in \mathbb{Q}[X, Y]$ such that $\langle X, Y \rangle = \langle f \rangle$. Make use of the degree function in $\mathbb{Q}[Y][X]$ with respect to X to reach a contradiction.

3. Give an example of a total ordering that is not a well ordering.

4. Why is a well order a total ordering?

5. Let $\leq$ be a term ordering on $\mathbb{N}^n$. Show that $a + c \leq b + d$ if $a \leq b$ and $c \leq d$, where $a, b, c, d \in \mathbb{N}^n$.

6. Suppose that $v \in \mathbb{R}^2$. Define the relation R_v on $\mathbb{N}^2$ by

   $v_1 R v_2 \Leftrightarrow v \cdot v_1 \leq v \cdot v_2$, where $\cdot$ refers to the usual scalar product.

   (i) Is $R_{(1,1)}$ a partial ordering?

   (ii) Is $R_{(1,\sqrt{2})}$ a partial ordering? Is it a term ordering?

   (iii) Is $R_{(-1,\sqrt{2})}$ a term ordering?

7. Prove that $\leq$ is reflexive, antisymmetric, transitive, total with $0 \leq v$, $v_1 \leq v_2 \Rightarrow (v_1 + v) \leq (v_2 + v)$ for every $v, v_1, v_2 \in \mathbb{N}^n$, where

   (i) $\leq = \leq_{\text{lex}}$,

   (ii) $\leq = \leq_{\text{grlex}}$.

8. Prove that $\leq_v$, defined in (5.1), is a term ordering.

9. Let $\alpha = (\alpha_1, \ldots, \alpha_n)$, $\beta = (\beta_1, \ldots, \beta_n) \in \mathbb{N}^n$. Define the relation R on $\mathbb{N}^n$ by

$$\alpha R \beta$$

if and only if $\alpha = \beta$ or $\sum_{i=1}^n \alpha_i < \sum_{i=1}^n \beta_i$ or $\sum_{i=1}^n \alpha_i = \sum_{i=1}^n \beta_i$ and the first coordinates α_i, β_i from the right that are different satisfy $\alpha_i > \beta_i$.

   (i) Show that R is a term ordering (thus R is reflexive, antisymmetric, transitive, total, with $0 R v$, $v_1 R v_2 \Rightarrow (v_1 + v) R (v_2 + v)$ for every $v, v_1, v_2 \in \mathbb{N}^n$).

   (ii) Show without using Lemma 5.1.5 or Corollary 5.1.7 that R is a well ordering.

   The relation R is called the graded reverse lexicographic ordering. Usually it is the "fastest" term ordering in Gröbner basis computations.

10. Show that the graded reverse lexicographic ordering of Exercise 5.9 is the same as the graded lexicographic ordering $\leq_{\text{grlex}}$ on $\mathbb{N}^2$. Give an example showing that the graded reverse lexicographic ordering is not the same as the graded lexicographic ordering on $\mathbb{N}^3$.

11. Let $f, g \in R[X_1, \ldots, X_n] \setminus \{0\}$ where R is a domain and let $\leq$ be a term ordering on R. Prove that

$$\text{in}_{\leq}(fg) = \text{in}_{\leq}(f) \, \text{in}_{\leq}(g).$$

12. Let $f, g \in R[X_1, \ldots, X_n] \setminus \{0\}$ where R is a domain and let $\leq$ be a term ordering on R. Prove that

$$\text{in}_\leq(f + g) \leq \max(\text{in}_\leq(f), \text{in}_\leq(g)).$$

13. Compute the remainder $f^{(f_1, f_2)}$, where

$$f = 1 + X^5 + X + Y + X^3Y + X^4Y + Y^2 + 2X^2Y^2 + XY^3$$

and $(f_1, f_2) = (X^3 + Y^2, X^2Y + 1)$, using the division algorithm (and the lexicographic ordering $X \geq Y$).
   (i) The same as above, but with (f_2, f_1).
   (ii) Compute the remainder $f^{(f_1, f_2)}$ assuming that $X \leq Y$.
14. Let $F = (X^2 + Y, X^2Y + 1) \subseteq k[X, Y]$, where k is a field and let $\leq$ be a term ordering on $k[X, Y]$. Show that F is not a Gröbner basis with respect to $\leq$.
15. Let $f \in I = \langle f_1, \ldots, f_m \rangle \subseteq k[X_1, \ldots, X_n]$ and suppose that $(f_1, \ldots, f_m)$ is a Gröbner basis (with respect to some term ordering $\leq$) for I. Prove that $(f_1, \ldots, f_m, f)$ is also a Gröbner basis for I.
16. Let $G = (g_1, \ldots, g_r) \subseteq k[X_1, \ldots, X_n]$ and $I = \langle g_1, \ldots, g_r \rangle$. Prove that G is a Gröbner basis if and only if $(f \in I \iff f^G = 0)$ for every $f \in I$.
17. Let R be a (commutative) ring and $a_1, \ldots, a_n, b_1, \ldots, b_n \in R$. Show that

$$a_1 a_2 \cdots a_n - b_1 b_2 \cdots b_n \in \langle a_1 - b_1, a_2 - b_2, \ldots, a_n - b_n \rangle. \quad (5.12)$$

Now assume that $f, f_1, \ldots, f_r \in k[X_1, \ldots, X_n]$ and that $f = P(f_1, \ldots, f_r)$ for a suitable polynomial $P \in k[T_1, \ldots, T_r]$. Apply (5.12) to prove that

$$f(X_1, \ldots, X_n) - P(T_1, \ldots, T_r) \in I,$$

where I is the ideal $\langle T_1 - f_1, \ldots, T_r - f_r \rangle$ in the polynomial ring

$$k[X_1, \ldots, X_n, T_1, \ldots, T_r].$$

18. Let $F = (X^2 + Y, X^2Y + 1) \subseteq \mathbb{Q}[X, Y]$ and $f = X^3Y + X^2Y + X + Y^2$. Consider the lexicographic ordering $\leq$ with $X \geq Y$.
   (i) Prove that $f \rightarrow_F 0$.
   (ii) Prove that $f^F \neq 0$ and $f^{F'} \neq 0$, where $F' = (X^2Y + 1, X^2 + Y)$.
19. Compute the reduced Gröbner basis of $(X^2 + Y, X + Y)$ using the lexicographic ordering $X \geq Y$.
20. Is $(X^2 + Y, X + Y)$ already a Gröbner basis with respect to some term ordering?

21. Decide whether $f = X^3Y + X^3 + X^2Y^3 - X^2Y + XY + X$ lies in the ideal $I = \langle X^2 + Y, \; X^2Y + 1 \rangle \subseteq k[X, Y]$. If so, find $a_1, a_2 \in k[X, Y]$ such that $f = a_1 f_1 + a_2 f_2$.

22. Let $I \subseteq k[X, Y, Z]$ denote the ideal $\langle X^2 - Y, \; Z^3 + Y^2 \rangle \subseteq k[X, Y, Z]$. Let $\leq$ denote the lexicographic ordering on $k[X, Y, Z]$ given by $X \geq Y \geq Z$.
    (i) Show that $(X^2 - Y, \; Z^3 + Y^2)$ is a reduced Gröbner basis with respect to $\leq$ for I.
    (ii) Show that $X^3 - XY + Y^2 + Z^4 + ZY^2 \notin I$.

23. Let I be the ideal $(f_1, f_2) = (X^2 + Y, \; X + Y) \subseteq k[X, Y]$.
    (i) Show that $f = X^2 + X^4 + X^2Y + X^3Y - Y^2 + XY^2 \in I$
    (ii) Compute $a_1, a_2 \in k[X, Y]$ such that $f = a_1 f_1 + a_2 f_2$.

24. Let $I \subseteq \mathbb{Q}[X, Y]$ denote the ideal $\langle X^2 + Y^2, \; X^3 + Y^3 \rangle \subseteq \mathbb{Q}[X, Y]$. Let $\leq$ denote the lexicographic ordering on $\mathbb{Q}[X, Y]$ given by $X \geq Y$.
    (i) Compute the S-polynomials $S_1 = S(X^2 + Y^2, X^3 + Y^3)$ and $S_2 = S(X^2 + Y^2, S_1)$ with respect to $\leq$ and show that $S_1, S_2 \in I$. Use this to prove that $Y^4 \in I$.
    (ii) Show that the reduced Gröbner basis for I with respect to $\leq$ is $(Y^4, \; XY^2 - Y^3, \; X^2 + Y^2)$.
    (iii) Show that $(X^2 + Y^2, \; X^3 + Y^3)$ cannot be a Gröbner basis for I for any term ordering.

25. Let R denote the ring $\mathbb{Q}[X, Y, S, T]$ and $\leq$ the lexicographic term ordering on R given by

$$X \geq Y \geq S \geq T.$$

Let I denote the ideal $R(S - X^2) + R(T - XY)$.
    (i) Show that the reduced Gröbner basis for I with respect to $\leq$ is

$$G = (X^2 - S, \; XY - T, \; XT - YS, \; Y^2S - T^2).$$

    (ii) Compute the remainder $Q = (X^4 + 2X^3Y)^G$. Show that $Q \in \mathbb{Q}[S, T]$ and that $X^4 + 2X^3Y = Q(X^2, XY)$.
    (iii) Let $f \in \mathbb{Q}[X, Y]$ and let Q denote the unique remainder f^G. Show that $f(X, Y) = Q(X^2, XY)$ if $Q \in \mathbb{Q}[S, T]$.

26. Let c denote the vector $(c_1, c_2) \in \mathbb{R}^2$ and let $c \cdot v = c_1 a + c_2 b$, where $v = (a, b) \in \mathbb{R}^2$. Define the relation R_c on $\mathbb{N}^2$ by

$$v_1 \; R_c \; v_2 \iff c \cdot v_1 \geq c \cdot v_2,$$

where $v_1, v_2 \in \mathbb{N}^2$.
    (i) Show that R_c is reflexive and transitive.
    (ii) Give an example showing that R_c is not necessarily antisymmetric.

(iii) Show that R_c is antisymmetric if $c_1/c_2 \notin \mathbb{Q}$, where $c_2 \neq 0$.

(iv) Let $c = (1, \sqrt{2})$. Show that R_c is a term ordering on $\mathbb{N}^2$. Compute the reduced Gröbner basis for the ideal $\langle X^2 + Y, X^2Y + 1 \rangle$ with respect to this term ordering (the term $X^m Y^n$ is identified with the vector $(m, n) \in \mathbb{N}^2$).

(v) Let $\leq \subseteq \mathbb{N}^2 \times \mathbb{N}^2$ denote the lexicographic term ordering on $\mathbb{N}^2$ given by $(1, 0) \geq (0, 1)$. Show that $\geq \neq R_c$ for every $c \in \mathbb{R}^2$.

27. Show that $X^2Z + Y \notin \langle XZ + Y^2, X + Y \rangle \subseteq \mathbb{Q}[X, Y, Z]$.

28. Let I denote the ideal generated by $X^2 + Y$ and $X^2Y + 1$ in $\mathbb{Q}[X, Y]$.

(i) Compute a Gröbner basis for I with respect to the lexicographic term ordering $\leq$, where $Y \geq X$.

(ii) Show that $Y^2 - 1$, $X^4 - 1 \in I$.

(iii) Let $\leq$ be an arbitrary term ordering. Prove that

$$(X^2 + Y, \ Y^2 - 1, \ X^4 - 1)$$

is a Gröbner basis for I with respect to $\leq$.

29. Show that the generators

$$I = \langle x^5 + y^3 + z^2 - 1, \ x^2 + y^2 + z - 1, \ x^6 + y^5 + z^3 - 1 \rangle$$

of Example 5.7.5 in fact form a Gröbner basis with respect to some term ordering (hint: construct a suitable weighted term ordering using (5.1)).

30. Let X be any subset of $k^n = k \times \cdots \times k$ (n times). Prove that

$$I(X) = \{f \in k[X_1, \ldots, X_n] \mid f(a_1, \ldots, a_n) = 0 \, \forall (a_1, \ldots, a_n) \in X\}$$

is an ideal in $k[X_1, \ldots, X_n]$. Show that $V(I(X)) \supseteq X$ and that $I(X) = I(V(I(X)))$.

31. Let $f_1, \ldots, f_m \in k[x_1, \ldots, x_n]$. Prove that

$$V(f_1, \ldots, f_m) = V(I),$$

where $I = \langle f_1, \ldots, f_m \rangle$.

32. Consider the ideal $I = \langle 5x + y + z - 17, \ x + y - z - 1, \ x + y + z - 9 \rangle \subseteq \mathbb{R}[x, y, z]$. Compute a Gröbner basis for I with respect to the lexicographic ordering $\leq$, where $x \geq y \geq z$. What is the relation to Gauss elimination when solving the system

$$5x + y + z = 17,$$
$$x + y - z = 1,$$
$$x + y + z = 9$$

of linear equations over $\mathbb{R}$?

33. **(HOF)** The following problem shows that every ideal has a finite generating set that is a Gröbner basis with respect to all term orderings. Such a generating set is called a universal Gröbner basis. Let $I \subseteq k[X_1, \ldots, X_n]$ be an ideal.

   (i) Show that there are only finitely many ideals generated by initial terms of elements in I. More precisely show that

   $$\{\mathrm{in}_\le(I) \mid \le \text{ term ordering on } k[X_1, \ldots, X_n]\}$$

   is a finite set. Where $\mathrm{in}_\le(I) = \langle \mathrm{in}_\le(f) \mid f \in I \backslash \{0\} \rangle$.

   (ii) Show that every ideal $I \subseteq k[X_1, \ldots, X_n]$ has a set of generators that is a Gröbner basis for every term ordering.

A Relations

In mathematical terms a relation on a set S is simply a subset $R \subseteq S \times S$. This definition is deceptively simple, but captures the real-world nature of relations remarkably. Of course, if one wants interesting mathematics one must restrict to relations with certain properties. The two most important types of relations in mathematics are equivalence relations and order relations.

A.1 Basic definitions and properties

Definition A.1.1 A *relation* R on a set S is a subset $R \subseteq S \times S$. We will write $x R y$ to mean $(x, y) \in R$.

Definition A.1.2 A relation R on S is *reflexive* if $x R x$ for every $x \in S$, *symmetric* if $x R y \Rightarrow y R x$ for every $x, y \in S$, *antisymmetric* if $x R y \wedge y R x \Rightarrow x = y$ for every $x, y \in S$ and *transitive* if $x R y \wedge y R z \Rightarrow x R z$ for every $x, y, z \in S$.

  (i) R is called an *equivalence relation* if it is reflexive, symmetric and transitive.
  (ii) R is called a *partial ordering* if it is reflexive, antisymmetric and transitive.

Example A.1.3 Recall the relation $\leq$ on $\mathbb{Z}$ from Chapter 1 given by $x \leq y \iff y - x \in \mathbb{N}$. Since $0 \in \mathbb{N}$, $\leq$ is reflexive. It is antisymmetric since if $x \in \mathbb{Z}$ and $x \in \mathbb{N}$, $-x \in \mathbb{N}$ then $x = 0$. It is transitive, as $x, y \in \mathbb{N}$ implies $x + y \in \mathbb{N}$. So $\leq$ is a partial ordering on $\mathbb{Z}$.

Example A.1.4 Let S be a set.
  (i) The relation $R = S \times S$ is reflexive, symmetric and transitive, but it is not antisymmetric if S contains more than one element.
  (ii) If the two relations $R_1, R_2 \subseteq S \times S$ both have one of the properties of Definition A.1.2 then the intersection $R_1 \cap R_2 \subseteq S \times S$ has the same property.

Example A.1.5   Let $I \subseteq R$ be an ideal in a commutative ring R. Then we define the relation (congruence modulo an ideal)

$$x \equiv y \,(\text{mod } I) \iff x - y \in I.$$

This relation is reflexive since $0 \in I$, symmetric since $x \in I \implies -x \in I$ and transitive since $x, y \in I \implies x + y \in I$. In short, congruence modulo I is an equivalence relation because I is a subgroup of R. As a special case we may take $I = d\mathbb{Z}$ in $\mathbb{Z}$. Then $x \equiv y \,(\text{mod } I)$ if and only if $x \equiv y \,(\text{mod } d)$. So congruence modulo an integer is an equivalence relation.

Example A.1.6   Suppose that R_1 and R_2 are relations on a set M. Then $R_1 \circ R_2$ is the relation on M given by $\{(x, z) \in M \times M \mid (x, y) \in R_1, (y, z) \in R_2$ for some $y \in M\}$. If R is a relation on M, we define R^n iteratively by $R^n = R \circ R^{n-1}$, where $n \in \mathbb{N}$ and $R^0 = M \times M$.

Let $S = \{1, 2, 3, 4\}$ and $R = \{(1, 2), (2, 4), (4, 2), (1, 1), (2, 3)\}$. Then R can be shown diagrammatically, the nodes correspond to the elements of S and the arrows to elements of the relation R. Below you will find diagrams of the relations R, R^2, R^3 and $R \cup R^2 \cup R^3$:

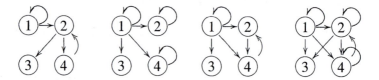

Notice that $R \cup R^2 \cup R^3$ is a transitive relation but R is not.

A.2 Equivalence relations

Let $\sim$ be an equivalence relation on a set S. Given $x \in S$, we let

$$[x] = \{s \in S \mid s \sim x\} \subseteq S.$$

This subset is called the *equivalence class* containing x and x is called a *representative* for $[x]$. The set of equivalence classes

$$\{[x] \mid x \in S\}$$

is denoted $S/\sim$.

Remark A.2.1   When dealing with equivalence relations, the symbol $\sim$ is often used instead of R ($x R y$ is denoted by $x \sim y$).

Example A.2.2   Let I be an ideal of a commutative ring R and let $\equiv$ denote equivalence modulo the ideal I (see Example A.1.5). Then $R/\equiv$ equals R/I.

You may have noticed that $[x]$ is defined as *the* equivalence class containing x. How can we be sure that there is just one equivalence class containing x? This is in fact a consequence of the following lemma.

Lemma A.2.3   *Let $\sim$ be an equivalence relation on S and $x, y \in S$. Then $[x] = [y]$ if and only if $x \sim y$.*

Proof.   Suppose that $[x] = [y]$. Then $x \in [x]$, since $\sim$ is reflexive. Therefore $x \in [y]$ and $x \sim y$. Let us prove that $[x] \subseteq [y]$ if $x \sim y$. Let $s \in [x]$. Then $s \sim x$ and since $x \sim y$ we get, by the transitivity of $\sim$, that $s \sim y$. Thus $s \in [y]$. Using that $x \sim y \implies y \sim x$, the same proof can be repeated to show that $[y] \subseteq [x]$ if $x \sim y$. $\square$

Corollary A.2.4   *In the notation of Lemma A.2.3, $[x] \cap [y] = \emptyset$ if $[x] \neq [y]$.*

Proof.   Suppose that $z \in [x] \cap [y] \neq \emptyset$. Then $z \sim x$ and $z \sim y$. By Lemma A.2.3 we have $[z] = [x]$ and $[z] = [y]$. Thus $[x] = [y]$. $\square$

Definition A.2.5   A *partition* of a set S is a collection $(S_i)_{i \in I}$ of subsets of S such that $\cup_{i \in I} S_i = S$ and $S_i \cap S_j = \emptyset$ if $i \neq j$.

The key property of equivalence relations is contained in the theorem below.

Theorem A.2.6   *Let S be a set with an equivalence relation $\sim$. Then the set of equivalence classes*

$$S/\sim = \{[x] \mid x \in S\}$$

is a partition of S. However, if $(S_i)_{i \in I}$ is a partition of S then we get an equivalence relation $\sim$ on S such that $S/\sim = (S_i)_{i \in I}$.

Proof.   We have already seen that equivalence classes are disjoint (Corollary A.2.4). We need to show that every element $x \in S$ is contained in an equivalence class. But this follows from the fact that $\sim$ is reflexive ($x \in [x]$). Suppose that $(S_i)_{i \in I}$ is a partition of S. We define $x \sim y \iff x, y \in S_i$ for some $i \in I$. Reflexivity follows from $\cup_{i \in I} S_i = S$. Symmetry is clear. Transitivity is implied by $S_i \cap S_j = \emptyset$ if $i \neq j$. If $x \in S_i$ then $S_i = [x]$. $\square$

Definition A.2.7   Let $\sim$ be an equivalence relation on a set S. Then the map

$$\pi : S \to S/\sim$$

given by $\pi(s) = [s]$ is called the canonical map.

Proposition A.2.8 *Let* $f : S \to M$ *be a map from a set S with an equivalence relation* $\sim$ *to a set M. If* $x \sim y \iff f(x) = f(y)$ *then there is an injective map*

$$\tilde{f} : S/\sim \to M$$

such that $f = \tilde{f} \circ \pi$, *where* π *is the canonical map.*

Proof. Let $\tilde{f}([x]) = f(x)$. This is a definition that depends on the choice of representative $x \in [x]$. But if $[y] = [x]$ then $y \sim x$ and $f(y) = f(x)$, so our map $\tilde{f}$ is actually well defined. It satisfies $f = \tilde{f} \circ \pi$ and is injective by construction.   □

A.2.1 Construction of the integers $\mathbb{Z}$

Even though it is somewhat formal, let us see how the concept of equivalence relations enables us to construct the integers $\mathbb{Z}$, given the natural numbers $\mathbb{N}$ with addition and multiplication. We look at pairs $(x, y) \in \mathbb{N} \times \mathbb{N}$. The pair (x, y) will be our candidate for the integer $x - y$. We introduce the relation $\sim$ given by

$$(x, y) \sim (x_1, y_1) \iff x + y_1 = y + x_1$$

on $\mathbb{N} \times \mathbb{N}$. You can easily check that this is an equivalence relation. The inspiration for $\sim$ is of course that $x - y = x_1 - y_1 \iff x + y_1 = y + x_1$. We define addition and multiplication as

$$(x, y) + (x_1, y_1) = (x + x_1, y + y_1),$$
$$(x, y)(x_1, y_1) = (xx_1 + yy_1, xy_1 + yx_1).$$

Now we may construct the integers as the equivalence classes

$$\mathbb{Z} = \mathbb{N} \times \mathbb{N}/\sim.$$

What about addition and multiplication? Is it safe to define

$$[(x, y)] + [(x_1, y_1)] = [(x + x_1, y + y_1)]?$$

Already at this point it is extremely important that you notice that a definition such as this is a problem. We use elements (x, y) in the equivalence classes $[(x, y)]$ to define $+$. What if we picked another element $(x', y') \in [(x, y)]$? Would the addition still give the same equivalence class?

The answer is yes. Here is a proof, which is typical of the procedure you must go through to ensure that an operation on equivalence classes is well defined. Suppose that $[(x, y)] = [(x', y')]$ and $[(x_1, y_1)] = [(x_1', y_1')]$. We must prove that $[(x + x_1, y + y_1)] = [(x' + x_1', y' + y_1')]$. Using the definition of $\sim$

we see that $x + x_1 + y' + y_1' = x + y' + x_1 + y_1' = y + x' + y_1 + x_1'$, showing that $(x + x_1, y + y_1) \sim (x' + x_1', y' + y_1')$, so that

$$[(x + x_1, y + y_1)] = [(x' + x_1', y' + y_1')].$$

The same proof (with a twist) works for multiplication. Notice that $[(m, n)] = [(m - n, 0)]$ if $m \geq n$ and that $[(m, n)] = [(0, n - m)]$ if $m \leq n$. Putting $-m = [(0, m)]$ for $m \in \mathbb{N}$ and identifying $n \in \mathbb{N}$ with $[(n, 0)]$ we have constructed the integers.

A.2.2 Construction of the rational numbers $\mathbb{Q}$

Now that we know the integers $\mathbb{Z} = \{\ldots, -3, -2, -1, 0, 1, 2, 3, \ldots\}$ and how to add and multiply them, how do we use equivalence relations to give a precise definition of the rational numbers $\mathbb{Q}$? A fraction is given by a numerator $a \in \mathbb{Z}$ and a denominator $b \in \mathbb{Z} \setminus \{0\}$, but then again this is not totally precise; two fractions, such as $\frac{1}{2}$ and $\frac{2}{4}$, may be the same even though they do not have the same numerators and denominators.

Suppose that we impose the relation $(a, s) \sim (b, t) \Leftrightarrow at = bs$ on the set $M = \mathbb{Z} \times (\mathbb{Z} \setminus \{0\})$. This is an equivalence relation and it mimics the everyday rule that you do not change a fraction if you multiply the numerator and denominator by the same non-zero number. Now define the subset

$$\frac{a}{s} = [(a, s)] = \{(b, t) \in M \mid (b, t) \sim (a, s)\} \subseteq M.$$

This subset is supposed to be our "fraction" a/s – of course no sane human being views a fraction as a huge set in this way, but read on! We have sorted out the infinite amount of identical fractions and made them into one object, just by naively putting things together that are considered the same. Now we finally define the rational numbers

$$\mathbb{Q} = M/\sim = \left\{ \frac{a}{b} \mid (a, b) \in M \right\}.$$

Does it make sense to add and multiply our fractions? Suppose that we simply define

$$\frac{a}{s}\frac{b}{t} = \frac{ab}{st}, \qquad \frac{a}{s} + \frac{b}{t} = \frac{at + bs}{st}.$$

As in the construction of $\mathbb{Z}$ given $\mathbb{N}$ one needs to check that the multiplication and addition is independent of the choice of representatives. This is left as an exercise.

A.3 Partial orderings

Example A.3.1 Here are some more examples of well known partial orderings.

(i) Let S be a set. Then inclusion $\subseteq$ is a partial ordering on the set of subsets of S.
(ii) Let R denote the relation on $\mathbb{N}$ given by $xRy \iff x \mid y$ for $x, y \in \mathbb{N}$. Then R is a partial ordering. But R is not a partial ordering considered as a relation on $\mathbb{Z}$ (why?).

An element $s \in S$ in a set with a partial ordering $\leq$ is said to be *minimal* if

$$x \leq s \implies x = s$$

for every $x \in S$. An element $t \in S$ is called a *first* element if

$$t \leq x$$

for every $x \in S$. Because of antisymmetry a first element has to be unique. A first element is a minimal element. What about the other way around? The answer is no: there is no reason why (s, x) should belong to the subset of $S \times S$ given by $\leq$. Here are some examples with several minimal elements.

Example A.3.2 Let

$$S = \{\{0\}, \{1\}, \{0, 1\}\}$$

be a set of subsets of $\{0, 1\}$. Then the inclusion of sets $\subseteq$ is a partial ordering on S and $\{0\}$ and $\{1\}$ are two different minimal elements of S.

Example A.3.3 Let $S = \mathbb{N} \setminus \{1\}$. The divisibility relation $xRy \iff x \mid y$ is a partial ordering on S. The fact that there are infinitely many primes in $\mathbb{N}$ tells us that R has infinitely many minimal elements.

Definition A.3.4 A partial ordering $\leq$ is called a *total ordering* if $x \leq y$ or $y \leq x$ for every $x, y \in S$.

An even finer condition is given by

Definition A.3.5 A partial ordering $\leq$ on a set S is called a *well ordering* if every non-empty subset $M \subseteq S$ has a first element $m \in M$.

Definition A.3.6 Let $\leq$ be a total ordering on a finite set M. Then we let $\max_{\leq} M$ denote the maximal element in M. Thus $x = \max_{\leq}(M)$ if and only if $x \in M$ and $x \geq y$ for every $y \in M$. Similarly we let $\min_{\leq}(M)$ denote the minimal element. When the ordering is implicit we drop the subscript and write max and min instead of $\max_{\leq}$ and $\min_{\leq}$.

Example A.3.7 The partial ordering $\leq$ on $\mathbb{Z}$ is not a well ordering, since $\mathbb{Z}$ does not have a first element. Every total ordering on a finite set is a well ordering. One of the surprising results of set theory is that there exists a well ordering on every set (can you construct one on $\mathbb{Z}$? $\mathbb{R}$?).

Lemma A.3.8 *Let S be a set with a well ordering $\leq$ and $F = \{s_1, s_2, \ldots\}$ a subset such that $s_1 \geq s_2 \geq s_3 \geq \ldots$. Then F is finite.*

Proof. Let s denote the smallest element of F. Since $s \in F$ this means that $s = s_N$ for some $N \in \mathbb{N}$. Since $s_N \geq s_i$ for $i > N$ this implies that $s_N = s_i$ for $i > N$. Therefore F is finite. $\qquad\square$

B Linear algebra

Vector spaces over the real numbers are familiar creatures. But the definition of a real vector space makes perfect sense when you replace the real numbers $\mathbb{R}$ by an arbitrary field F. The crucial thing is that given a non-zero $x \in F$ there is a $y \in F$ such that $xy = 1$.

Definition B.0.9 A vector space V over a field F is an abelian group $(V, +)$ with neutral element 0 and a (scalar) multiplication $F \times V \to V$ denoted $(a, v) \mapsto av$ such that

(i) $(ab)v = a(bv)$
(ii) $1v = v$
(iii) $(a + b)v = av + bv$
(iv) $a(v + w) = av + aw$

for every $a, b \in F$ and every $v, w \in V$.

A subspace of V is a subgroup $W \subseteq V$ such that $av \in W$ if $a \in F$ and $v \in W$. A group homomorphism $\varphi : V \to W$ between vector spaces V and W over a field F is called a linear map if $\varphi(av) = a\varphi(v)$ where $a \in F$ and $v \in V$.

Let $\varphi : V \to W$ be a linear map. The subset $\mathrm{Ker}\,(\varphi) = \{v \in V \mid \varphi(v) = 0\} \subseteq V$ is called the kernel of φ and $\mathrm{Im}(\varphi) = \{\varphi(v) \mid v \in V\} \subseteq W$ is called the image of φ. Both are subspaces. Let $V' \subseteq V$ be a subspace; then the quotient group $V/V' = \{v + V' \mid v \in V\}$ is a vector space through the (well defined) scalar multiplication given by $a(v + V') = av + V'$. By Theorem 2.5.1 we have a group isomorphism

$$\tilde{\varphi} : V/\mathrm{Ker}\,(\varphi) \to \mathrm{Im}(\varphi),$$

where $\varphi : V \to W$ is a linear map. This group isomorphism is also a linear map.

A *vector space* V over a field F is called finitely generated if there exists a finite set of vectors $v_1, \ldots, v_n \in V$ such that every $v \in V$ can be written as a *linear combination* $v = a_1 v_1 + \cdots + a_n v_n$ for suitable $a_1, \ldots, a_n \in F$. Such a set of vectors is called a (finite) generating set for V. We will assume that vector spaces are finitely generated.

Example B.0.10   Let F be a field. Then

$$V = F^n = F \times \cdots \times F$$

is a vector space resembling $\mathbb{R}^n$. The multiplication $F \times V \to V$ is given by

$$a(v_1, \ldots, v_n) = (av_1, \ldots, av_n),$$

where $a \in V$ and $(v_1, \ldots, v_n) \in V$. The vectors $e_1 = (1, 0, \ldots, 0)$, $e_2 = (0, 1, 0, \ldots, 0)$, $\ldots$, $e_n = (0, 0, \ldots, 1) \in V$ form a generating set for V.

B.1   Linear independence

Definition B.1.1   Let V be a vector space over a field F. A set of vectors $v_1, \ldots, v_n$ is called *linearly independent* if

$$a_1 v_1 + \cdots + a_n v_n = 0,$$

where $a_1, \ldots, a_n \in F$ implies that $a_1 = \cdots = a_n = 0$.

We now prove what is known as the *Steinitz exchange lemma*.

Lemma B.1.2   *Let V be a vector space over a field F, $w_1, \ldots, w_m$ a linearly independent set of vectors and $v_1, \ldots, v_n$ a generating set for V. Then $m \leq n$ and $w_1, \ldots, w_m, v'_{m+1}, \ldots, v'_n$ gives a generating set for V, where $v'_{m+1}, \ldots, v'_n \in \{v_1, \ldots, v_n\}$.*

Proof.   We may assume by rearranging $v_1, \ldots, v_n$ that $w_1 = a_1 v_1 + \cdots + a_n v_n$, with $a_1 \neq 0$. This gives that v_1 can be written as a linear combination of $w_1, v_2, \ldots, v_n$. Therefore $w_1, v_2, \ldots, v_n$ is a generating set for V. We continue this procedure with w_2. Write $w_2 = a_1 w_1 + a_2 v_2 + \cdots + a_n v_n$. Here we must have $a_i \neq 0$ for some $i > 1$, otherwise w_1 and w_2 would not be linearly independent. Assume that $a_2 \neq 0$. In the same way as before $w_1, w_2, v_3, \ldots, v_n$ is a generating set. Proceeding like this we cannot exceed the nth vector v_n; This would contradict that $w_1, \ldots, w_m$ is a linearly independent set of vectors. Thus $m \leq n$ and in the process we have also shown that $w_1, \ldots, w_m, v'_{m+1}, \ldots, v'_n$ is a generating set for V, where $v'_{m+1}, \ldots, v'_n \in \{v_1, \ldots, v_n\}$. $\square$

Definition B.1.3   A basis for a vector space V is a linearly independent generating set for V.

Proposition B.1.4 *A (finitely generated) vector space V over a field F has a basis. More precisely, a minimal generating set for V is linearly independent.*

Proof. Let $v_1, \ldots, v_n \in V$ be a minimal generating set for V. This means that if we exclude any of $v_1, \ldots, v_n$ we are left with a set of vectors that is not a generating set. We wish to prove that $v_1, \ldots, v_n$ is a linearly independent set. If not, we would have $a_1, \ldots, a_n \in F$, not all zero, such that

$$a_1 v_1 + \cdots + a_n v_n = 0.$$

We may assume that $a_1 \neq 0$. This means that

$$v_1 = -a_1^{-1} a_2 v_2 - \cdots - a_1^{-1} a_n v_n.$$

Therefore $v_2, \ldots, v_n$ is a generating set, contradicting that $v_1, \ldots, v_n$ is a minimal generating set. □

Proposition B.1.5 *If $v_1, \ldots, v_m$ and $w_1, \ldots, w_n$ are two bases of a vector space then $m = n$.*

Proof. This follows from Lemma B.1.2. □

B.2 Dimension

Definition B.2.1 Let V be a vector space over a field F. The *dimension* $\dim_F V$ of V over F is the number of elements in a basis of V.

Proposition B.2.2 *Let V be a vector space over a field F and $W \subseteq V$ a subspace of V. If $\dim_F W = \dim_F V$ then $W = V$.*

Proof. Let $n = \dim_F W = \dim_F V$. Suppose that $w_1, \ldots, w_n$ is a basis for W and $v_1, \ldots, v_n$ a basis for V. Then we may use Lemma B.1.2 to conclude that $w_1, \ldots, w_n$ is also a basis for V. Thus $W = V$. □

Proposition B.2.3 *Let V be a vector space over a field F and $W \subseteq V$ a subspace of V. Then*

(i) $\dim_F V/W = \dim_F V - \dim_F W$,
(ii) $\dim_F V + W = \dim_F V + \dim_F W - \dim_F V \cap W$,
(iii) $\dim_F \mathrm{Ker}(\varphi) + \dim_F \mathrm{Im}(\varphi) = \dim_F V$ *where $\varphi : V \to W$ is a linear map.*

Proof. If $w_1, \ldots, w_r$ is a basis for W and $v_1, \ldots, v_n$ a basis for V then we may assume that $w_1, \ldots, w_r, v_{r+1}, \ldots, v_n$ is a basis for V by Lemma B.1.2. It is easy to verify that $v_{r+1} + W, \ldots, v_n + W$ is a basis for V/W. This proves the first formula. To prove the second formula recall that $V + W$ is the subspace defined as $\{v + w \mid v \in V,\ w \in W\}$ and that V and W are subspaces of $V + W$. The composed map

$$\psi : V \to V + W \to V + W/W$$

is given by $\psi(v) = v + W$. It is linear and surjective and $\operatorname{Ker}(\psi) = V \cap W$, since $\psi(v) = v + W = W$ if and only if $v \in W$. Therefore

$$V/V \cap W \cong V + W/W.$$

This shows that $\dim_F V/V \cap W = \dim_F V + W/W$, and (ii) follows from (i). Use the isomorphism

$$V/\operatorname{Ker}(\varphi) \cong \operatorname{Im}(\varphi)$$

to deduce the formula in (iii). □

If $\varphi : V \to W$ is a linear map, $v_1, \ldots, v_m$ a basis of V and $w_1, \ldots, w_n$ a basis of W then

$$\varphi(v_1) = a_{11}w_1 + \cdots + a_{n1}w_n.$$

$$\vdots$$

$$\varphi(v_m) = a_{1m}w_1 + \cdots + a_{nm}w_n,$$

If $v = x_1 v_1 + \cdots + x_m v_m \in V$, where $x_1, \ldots, x_m \in V$, we see that

$$\varphi(v) = x_1 \varphi(v_1) + \cdots + x_m \varphi(v_m).$$

Thus

$$\begin{pmatrix} y_1 \\ \vdots \\ y_n \end{pmatrix} = \begin{pmatrix} a_{11} & \cdots & a_{1m} \\ \vdots & \ddots & \vdots \\ a_{n1} & \cdots & a_{nm} \end{pmatrix} \begin{pmatrix} x_1 \\ \vdots \\ x_m \end{pmatrix},$$

where $\varphi(v) = y_1 w_1 + \cdots + y_n w_n$.

Example B.2.4 Let F be a field and $f = X^n + a_{n-1}X^{n-1} + \cdots + a_1 X + a_0 \in F[X]$ a polynomial of degree $n \geq 1$. Then

$$R = F[X]/\langle f \rangle$$

is a vector space over F with basis $1, \alpha, \ldots, \alpha^{n-1}$, where $\alpha = [X] \in R$. Multiplication by α is a linear map $\varphi : R \to R$. The matrix of φ with respect to the basis $1, \alpha, \ldots, \alpha^{n-1}$ is

$$\begin{pmatrix} 0 & 0 \ldots 0 & -a_0 \\ 1 & 0 \ldots 0 & -a_1 \\ 0 & 1 \ldots 0 & -a_2 \\ \vdots & \vdots \ddots \vdots & \vdots \\ 0 & 0 \ldots 1 & -a_{n-1} \end{pmatrix}.$$

The above facts are consequences of Proposition 4.6.7.

References

[1] R. Alford, A. Granville and C. Pomerance, There are infinitely many Carmichael numbers, *Annals of Mathematics* **139** (1994), 703–722.

[2] A. Archer, A modern treatment of the 15-puzzle, *American Mathematical Monthly* **106** (1999), 793–799.

[3] E. Berlekamp, Factoring polynomials over finite fields, *Bell System Technical Journal* **46** (1967), 1853–1859.

[4] R. Courant and H. Robbins, *What is Mathematics?*, Oxford University Press, 1941.

[5] R. E. Crandall, *Mathematica for the Sciences*, Addison Wesley, 1991.

[6] M. Demazure, *Cours d'algèbre*, Cassini, 1997.

[7] L. E. Dickson, Finiteness of the odd perfect and primitive abundant numbers with *n* distinct prime factors, *American Journal of Mathematics* **35** (1913), 413–422.

[8] W. Diffie and M. E. Hellman, New directions in cryptography, *IEEE Transactions on Information Theory* **22**, (1976), 644–654.

[9] W. Feit and J. Thompson, Solvability of groups of odd order, *Pacific Journal of Mathematics* **13** (1963), 775–1029.

[10] M. Gardner, *Mathematical Puzzles of SAM LOYD*, Dover Publications, 1959.

[11] C. F. Gauss, *Disquisitiones Arithmeticae*, Springer Verlag, New York, 1993, English translation by Arthur A. Clarke.

[12] G. H. Hardy and E. M. Wright, *An Introduction to the Theory of Numbers*, Oxford University Press, 1945.

[13] D. Hilbert, Über die Theorie der algebraischen Formen, *Mathematische Annalen* **36** (1890), 473–534.

[14] K. Ireland and M. Rosen, *A Classical Introduction to Modern Number Theory*, Springer Verlag, New York, 1982.

[15] J. L. W. Jensen, Om talteoretiske Egenskaber ved de Bernoulliske tal, *Nyt Tidsskrift for Mathematik* **26** (1915), 73–83.

[16] D. E. Knuth, *The Art of Computer Programming*, second edition, vol. II: *Seminumerical Algorithms*, Addison Wesley, New York, 1981.

[17] N. Koblitz, *A Course in Number Theory and Cryptography*, Springer Verlag, New York, 1987.

[18] O. Ore, *Number Theory and its History*, Dover Publications, New York, 1948.

[19] C. Pomerance, A tale of two sieves, *Notices of the American Mathematical Society* **43** (1996), 1473–1485.

[20] M. Rabin, Probabilistic algorithm for testing primality, *Journal of Number Theory* **12** (1980), 128–138.

[21] D. Redfern, *The Maple Handbook*, vol. Maple V Release 4, Springer Verlag, 1996.

[22] R. Rivest and A. Shamir and L. Adleman, A method for obtaining digital signatures and public-key cryptosystems, *Communications of the ACM* **21** (1978), 120–126.

[23] B. Sturmfels, *Gröbner Bases and Convex Polytopes*, American Mathematical Society, Providence, 1996.

[24] L. Sylow, Théorèmes sur les groupes de substitutions, *Mathematische Annalen* **5** (1872), 584–594.

[25] S. Wagon, The Euclidean algorithm strikes again, *American Mathematical Monthly* **97** (1990) 125–129.

[26] A. Wiles, Modular elliptic curves and Fermat's last theorem, *Annals of Mathematics* **142** (1995), 443–551.

Index